Managing Construction AND Infrastructure
in the 21st Century Bureau of Reclamation

Committee on Organizing to Manage Construction and Infrastructure in the
21st Century Bureau of Reclamation

Board on Infrastructure and the Constructed Environment

Division on Engineering and Physical Sciences

NATIONAL RESEARCH COUNCIL
OF THE NATIONAL ACADEMIES

THE NATIONAL ACADEMIES PRESS
Washington, D.C.
www.nap.edu

THE NATIONAL ACADEMIES PRESS   500 Fifth Street, N.W.   Washington, DC 20001

NOTICE: The project that is the subject of this report was approved by the Governing Board of the National Research Council, whose members are drawn from the councils of the National Academy of Sciences, the National Academy of Engineering, and the Institute of Medicine. The members of the committee responsible for the report were chosen for their special competences and with regard for appropriate balance.

This study was supported by Contract Number 04CS811007 between the U.S. Department of the Interior and the National Academy of Sciences. Any opinions, findings, conclusions, or recommendations expressed in this publication are those of the author(s) and do not necessarily reflect the views of the organizations or agencies that provided support for the project.

International Standard Book Number 0-309-10035-6

Additional copies of this report are available from the National Academies Press, 500 Fifth Street, N.W., Lockbox 285, Washington, DC 20055; (800) 624-6242 or (202) 334-3313 (in the Washington metropolitan area); Internet, http://www.nap.edu.

Cover photographs from top to bottom: Parker Dam (from U.S. Bureau of Reclamation); deflector at Tracy fish screen (from San Luis and Delta Mendota Canal Authority); Flat Iron Power Plant and Pumping Station (from U.S. Bureau of Reclamation); Provo River restoration (from Utah Reclamation Mitigation and Conservation Commission); and Boise River Diversion Dam (from U.S. Bureau of Reclamation).

Copyright 2006 by the National Academy of Sciences. All rights reserved.

Printed in the United States of America

THE NATIONAL ACADEMIES
Advisers to the Nation on Science, Engineering, and Medicine

The **National Academy of Sciences** is a private, nonprofit, self-perpetuating society of distinguished scholars engaged in scientific and engineering research, dedicated to the furtherance of science and technology and to their use for the general welfare. Upon the authority of the charter granted to it by the Congress in 1863, the Academy has a mandate that requires it to advise the federal government on scientific and technical matters. Dr. Ralph J. Cicerone is president of the National Academy of Sciences.

The **National Academy of Engineering** was established in 1964, under the charter of the National Academy of Sciences, as a parallel organization of outstanding engineers. It is autonomous in its administration and in the selection of its members, sharing with the National Academy of Sciences the responsibility for advising the federal government. The National Academy of Engineering also sponsors engineering programs aimed at meeting national needs, encourages education and research, and recognizes the superior achievements of engineers. Dr. Wm. A. Wulf is president of the National Academy of Engineering.

The **Institute of Medicine** was established in 1970 by the National Academy of Sciences to secure the services of eminent members of appropriate professions in the examination of policy matters pertaining to the health of the public. The Institute acts under the responsibility given to the National Academy of Sciences by its congressional charter to be an adviser to the federal government and, upon its own initiative, to identify issues of medical care, research, and education. Dr. Harvey V. Fineberg is president of the Institute of Medicine.

The **National Research Council** was organized by the National Academy of Sciences in 1916 to associate the broad community of science and technology with the Academy's purposes of furthering knowledge and advising the federal government. Functioning in accordance with general policies determined by the Academy, the Council has become the principal operating agency of both the National Academy of Sciences and the National Academy of Engineering in providing services to the government, the public, and the scientific and engineering communities. The Council is administered jointly by both Academies and the Institute of Medicine. Dr. Ralph J. Cicerone and Dr. Wm. A. Wulf are chair and vice chair, respectively, of the National Research Council.

www.national-academies.org

COMMITTEE ON ORGANIZING TO MANAGE CONSTRUCTION AND INFRASTRUCTURE IN THE 21ST CENTURY BUREAU OF RECLAMATION

JAMES K. MITCHELL, *Chair*, Virginia Polytechnic Institute and University, Blacksburg, Virginia
PATRICK R. ATKINS, Alcoa, New York, New York
ALLAN V. BURMAN, Jefferson Solutions, Washington, D.C.
TIMOTHY J. CONNOLLY, HDR Engineering, Inc., Omaha, Nebraska
LLOYD A. DUSCHA, U.S. Army Corps of Engineers (retired), Reston, Virginia
G. BRIAN ESTES, Consulting Engineer, Williamsburg, Virginia
MARTHA S. FELDMAN, University of California, Irvine
DARRELL G. FONTANE, Colorado State University, Fort Collins
SAMMIE D. GUY, Consulting Engineer, Falls Church, Virginia
L. MICHAEL KAAS, Consulting Engineer, Arlington, Virginia
CHARLES I. McGINNIS, U.S. Army Corps of Engineers (retired), Charlottesville, Virginia
ROGER K. PATTERSON, Nebraska Department of Natural Resources (retired), Lincoln

Staff

LYNDA L. STANLEY, Director, Board on Infrastructure and the Constructed Environment
MICHAEL D. COHN, Program Officer
DANA CAINES, Financial Associate
PAT WILLIAMS, Senior Project Assistant

BOARD ON INFRASTRUCTURE AND THE CONSTRUCTED ENVIRONMENT

HENRY HATCH, *Chair*, U.S. Army Corps of Engineers (retired), Oakton, Virginia
MASSOUD AMIN, University of Minnesota, Minneapolis
REGINALD DesROCHES, Georgia Institute of Technology, Atlanta
DENNIS DUNNE, Consultant, Scottsdale, Arizona
PAUL FISETTE, University of Massachusetts, Amherst
LUCIA GARSYS, Hillsborough County, Florida
WILLIAM HANSMIRE, Parsons Brinckerhoff Quade & Douglas, Detroit, Michigan
THEODORE C. KENNEDY, BE&K, Inc., Birmingham, Alabama
SUE McNEIL, University of Delaware, Wilmington
DEREK PARKER, Anshen+Allen, San Francisco, California
HENRY SCHWARTZ, JR., Washington University, St. Louis, Missouri
WILLIAM WALLACE, Rensselaer Polytechnic Institute, Troy, New York
CRAIG ZIMRING, Georgia Institute of Technology, Atlanta

Staff

LYNDA STANLEY, Director
MICHAEL D. COHN, Program Officer
KEVIN M. LEWIS, Program Officer
DANA CAINES, Financial Associate
PAT WILLIAMS, Senior Project Assistant

Preface

The Bureau of Reclamation (Reclamation) has a long history of accomplishments, and through this study and other efforts is preparing to continue its successful record of providing water and hydroelectric power in the western United States. Successful accomplishment of Reclamation's current mission in the twenty-first century—to manage, develop, and protect water and related resources in an environmentally and economically sound manner in the interest of the American public—is impacted, and in some cases dominated, by several new realities that are discussed in this report, including environmental factors, American Indian water rights, rural water needs, urbanization, increasing budget constraints, a broader set of stakeholders, an aging workforce, and an aging infrastructure.

The committee was not asked to assess the robustness of Reclamation in the face of extraordinary events, but the recent disasters caused by the hurricanes in the Gulf Coast region have brought that question to the attention of the committee. In the short term, the dispersed geography, decentralized line organization, and centralized service center of Reclamation should allow it to respond to localized events effectively. Over the long term, the bureau has exhibited its ability to deal with disasters, as shown in its response to the failure of Teton Dam in 1976. That event led to the creation of a robust safety of dams program, risk analysis and design review procedures, and an active effort to learn from past experience. The committee also observed active efforts to plan responses to developing problems caused by persistent drought conditions in the West. If faced with unexpected catastrophic events, Reclamation can be expected, in the committee's opinion, to rise to the challenge.

All the committee members, whose abbreviated biographies are given in Appendix A, contributed enormously to the successful completion of the study. They provided diverse expertise and a wealth of knowledge and experience in relevant disciplines and topics: organizational, construction, and operational history of the bureau, water resources engineering and planning, government policies and procedures, large organization management, human resources issues, and political considerations, among others. Each member brought a creative and fresh perspective to the study and participated in the drafting of the report and in the crafting of the several findings and recommendations. It has been a pleasure and excellent learning experience working with all of them.

An important element in the committee's ability to complete its assigned tasks was the support and participation of the bureau. The committee appreciates the cooperation and support of John Keys III, commissioner, the assistance provided throughout the study by Fred Ore, deputy director of operations, and N. John Harb, manager, and the scores of managers and personnel in the Denver, regional, and area offices who took time from their busy schedules to brief the committee and candidly discuss Reclamation's challenges and opportunities. The committee also appreciates the contributions of Reclamation's water and power customers and their representative organizations, which provided a perspective on the bureau that was critical to the committee's understanding of the factors that influence its facility and infrastructure tasks.

The committee was supported and guided in its work by study director Michael Cohn, program officer, Board on Infrastructure and the Constructed Environment (BICE). Mike's dedication to the tasks and support for the committee is a key factor in the success of this study. We are also greatly indebted to Lynda Stanley, director, BICE, for her insights and suggestions.

The committee appreciates the opportunity to address an issue of importance to the future success of the Bureau of Reclamation's mission in meeting water and hydroelectric power needs in the western United States in an environmentally sensitive and economical manner.

James K. Mitchell

Chair, Committee on Organizing
to Manage Construction and
Infrastructure in the 21st Century
Bureau of Reclamation

Acknowledgment of Reviewers

This report has been reviewed in draft form by individuals chosen for their diverse perspectives and technical expertise, in accordance with procedures approved by the NRC's Report Review Committee. The purpose of this independent review is to provide candid and critical comments that will assist the institution in making its published report as sound as possible and to ensure that the report meets institutional standards for objectivity, evidence, and responsiveness to the study charge. The review comments and draft manuscript remain confidential to protect the integrity of the deliberative process. We wish to thank the following individuals for their review of this report:

John T. Christian, Consulting Engineer,
David W. Fowler, University of Texas at Austin,
Gerald E. Galloway, University of Maryland,
Lawrence J. MacDonnell, Porzak, Browning & Bushong,
Peter Marshall, Burns & Roe Services,
Robert S. O'Neil, Parsons Transportation Group (retired), and
Karlene H. Roberts, University of California, Berkeley.

Although the reviewers listed have provided many constructive comments and suggestions, they were not asked to endorse the conclusions or recommendations, nor did they see the final draft of the report before its release. The review of this report was overseen by Richard N. Wright, Building and Fire Research Laboratory, National Institute of Standards and Technology (retired). Appointed by the National Research Council, he was responsible for making certain that an independent examination of this report was carried out in accordance with institutional procedures and that all review comments were carefully considered. Responsibility for the final content of this report rests entirely with the authoring committee and the institution.

Contents

EXECUTIVE SUMMARY 1

1 INTRODUCTION 15
Background, 15
Summary of Authorizing Legislation, 15
Mission, 17
Statement of Task, 18
Organization of the Report, 19
References, 21

2 REQUIREMENTS FOR THE TWENTY-FIRST CENTURY 22
Introduction, 22
Facility and Infrastructure Assets, 23
Workload, 26
Management Policies and Procedures, 33
Decision-Making Procedures, 35
Organizational Configuration, 36
References, 46

3 GOOD PRACTICE TOOLS AND TECHNIQUES 48
Introduction, 48
Roundtable of Organizations with Similar Missions, 48
Policies and Procedures, 54
Acquisition and Contracting Practices, 56

Project Conception, Development, and Execution Practices, 60
Customer and Stakeholder Relations, 66
Application of Metrics, Audits, and Reviews, 68
Planning and Budgeting, 68
References, 69

4 WORKFORCE AND HUMAN RESOURCES 71
Introduction, 71
Workforce Planning, 72
Strategic Direction, 73
Supply Analysis, 77
Demand Analysis, 79
Gap Analysis, 80
Solutions and Implementation, 82
Evaluation, 86
References, 86

5 ALTERNATIVE SCENARIOS FOR FUTURE INFRASTRUCTURE MANAGEMENT 88
Introduction, 88
Scenario 1: Centrally Located Project Management Organization, 89
Scenario 2: Outsourced Operations and Maintenance, 91
Scenario 3: Federal Funding and Local Execution, 91
Conclusion, 92
Reference, 93

6 CONCLUSIONS, FINDINGS, AND RECOMMENDATIONS 94
Introduction, 94
Factors Impacting the Management of Construction and Infrastructure, 95
Capabilities for the Management of Construction and Infrastructure, 104
Alternative Scenarios for Future Infrastructure Management, 107

APPENDIXES

A BIOGRAPHIES OF COMMITTEE MEMBERS 111
B BRIEFINGS TO THE COMMITTEE AND DISCUSSIONS 119
C GOOD PRACTICE TOOLS AND TECHNIQUES ROUNDTABLE 128

BOARD ON INFRASTRUCTURE AND THE CONSTRUCTED ENVIRONMENT 138

Acronyms and Abbreviations

ALP   Animas–La Plata Project

BRC   Budget Review Committee

CALFED   CALFED Bay–Delta Program
CBT   Colorado–Big Thompson project
CCE   construction cost estimate
CFR   comprehensive facility review
CII   Construction Industry Institute
COTR   contracting officer's technical representative
CPORT   Commissioner's Program and Organization Review Team
CVP   Central Valley Project

DEC   Design, Estimating, and Construction Office
DOE   Department of Energy
DOI   Department of the Interior
DSIS   Dam Safety Information System
DSO   Dam Safety Office
DWR   California Department of Water Resources

EIA   environmental impact assessment
ESA   Endangered Species Act

FAR	federal acquisition regulations	
FFC	Federal Facilities Council	
GSA	General Services Administration	
IDIQ	indefinite delivery/indefinite quantity	
IDP	individual development plan	
KSAs	knowledge, skills, and abilities	
M&I	municipal and industrial	
MSCP	Multi-Species Conservation Program	
NCWCD	Northern Colorado Water Conservancy District	
NOAA	National Oceanic and Atmospheric Administration	
NWRA	National Water Resources Association	
O&M	operations and maintenance	
OMB	Office of Management and Budget	
OPP	Office of Procurement Policy	
OPPS	Office of Program and Policy Services	
PBSA	performance-based services acquisition	
PCE	project cost estimate	
PFR	periodic facility review	
PMP	project management plan	
PMT	project management team	
PMTS	Policy Management and Technical Services	
QA/QC	quality assurance and quality control	
R&D	research and development	
RAX	replacement, addition, and exceptional maintenance	
RDCCT	Reclamation Design and Construction Coordination Team	
Reclamation	U.S. Bureau of Reclamation	
SABER	simplified acquisition of basic engineering requirements	
SEED	safety evaluation of existing dams	
SES	Senior Executive Service	
SOD	Safety of Dams (program)	
SSLE	Security, Safety, and Law Enforcement	
SWP	state water project	

TSC	Technical Service Center
TVA	Tennessee Valley Authority
USACE	U.S. Army Corps of Engineers
USBR	U.S. Bureau of Reclamation
USGS	U.S. Geological Survey
WAPA	Western Area Power Administration (DOE)
WARSMP	Watershed and River System Management Program
WQIC	Water Quality Improvement Center

Executive Summary

In the more than 100 years since President Theodore Roosevelt signed the Reclamation Act in 1902, the U.S. Bureau of Reclamation (Reclamation) has compiled an enviable record, and it can take justifiable pride in having brought water and electrical power to the arid regions of the 17 western states. Over the course of the twentieth century, Reclamation participated in such monumental undertakings as the construction of the Hoover and Grand Coulee dams as well as the development of many other dams, reservoirs, hydroelectric plants, and massive irrigation systems. These facilities and infrastructure systems have provided the water and power that enabled the development and growth of agriculture, industry, commerce, cities, and towns in the West.

Reclamation is now the largest water wholesaler in the country, providing municipal and industrial water to more than 31 million people and irrigation water for 10 million acres that produce 60 percent of the nation's vegetables and 25 percent of its nuts and fruits (USBR, 2005). It is the second-largest producer of hydroelectric power, generating 42 billion kilowatt-hours of electricity annually. The bureau also partners in the management of more than 300 recreation sites.

Major water and power systems are now in place, and relatively few large new projects are anticipated. As a consequence, the bureau's focus and workload have shifted from building infrastructure to operating, maintaining, repairing, and modernizing it, and from constructing dams to evaluating dam safety, mitigating the risk of dam failure, and addressing environmental issues. Reclamation's budget has been level while at the same time the cost of maintaining and repairing existing infrastructure is rising, in part owing to aging facilities, normal wear and tear, and increased stakeholder attention to environmental issues.

As the West has grown, the demand for water and power has also grown. At the same time, laws have been enacted to protect ecosystems and mitigate the impacts of development on fish and wildlife. These events and others have created an operating environment in which water rights issues, water and power user interests, environmental concerns, American Indian tribal rights, and other considerations play a more and more important role in decision making, project management, and customer and stakeholder relations. Reclamation works with a broad range of stakeholders, some of whom have opposing objectives and values.

As part of the sustained effort to reinvent government, Congress has mandated that all federal executive agencies become more customer-service-oriented, more cost-effective, and more accountable for the results of their programs. Congress has also enacted legislation that expands agencies' options for procuring and delivering goods and services and, in some instances, for financing projects. Additionally, initiatives have been undertaken to downsize the federal workforce and outsource to the private sector work traditionally conducted by government employees. In response to these initiatives, Reclamation reorganized in the mid-1990s in order to streamline its management structure and eliminated many senior management positions. Services were centralized for the sake of efficiency and economy, and operational authority was delegated to field offices. Centralized oversight was loosened dramatically as mandatory procedural directives and standards were eliminated to allow greater flexibility in decision making and to empower field managers and staff to work more closely with Reclamation's customers. Reclamation also instituted some measures to manage its services through fee-for-service and cost recovery programs.

In the coming decades, population and development in the West are projected to continue to increase. As growth occurs, more land in agricultural use is likely to be used for municipal and industrial development. These changes will spur demand for more water and power resources, and that demand may outstrip the supply. Reclamation will be challenged to find ways to manage water and power so that it can meet future demand. Reclamation's tasks will involve water conservation; dam safety; expanding the existing capacity for desalination, water storage, and transmission; enhancing the recovery of endangered species and environmental quality; constructing new facilities to implement American Indian water rights settlements; removing dams; and operating, maintaining, repairing, and improving existing facilities. These changing and expanding requirements will occur at a time when the personnel with the most technical expertise and the best institutional memory regarding specific projects and stakeholders will be eligible to retire.

Reclamation has recognized the challenge for the twenty-first century and the necessity of making the transition from a construction organization to a resources management organization. Although Reclamation's mission continues to be the effective management of power and water in ways that protect the health, safety, and welfare of the American public and are environmentally and economically sound, achieving these objectives is a dynamic, complex, and uncertain matter.

OVERVIEW OF THE STUDY

At the request of Department of the Interior, the National Research Council (NRC) appointed the Committee on Organizing to Manage Construction and Infrastructure in the 21st Century Bureau of Reclamation, a group of experts from the public and private sectors and academia to advise Reclamation and the department on the "appropriate organizational, management, and resource configurations to meet its construction, maintenance, and infrastructure requirements for its missions of the 21st century." The full statement of task is presented in Chapter 1.

To accomplish its tasks the committee met as a whole four times from February to August 2005 and conducted small-group site visits to offices and projects in each of the five Reclamation regions. The committee received briefings from and had discussions with Reclamation representatives, Reclamation's customers and other stakeholders, and representatives of organizations with missions similar to Reclamation's.

During the course of this study the committee observed that the five Reclamation regions have different organizational structures, capabilities, and workloads. In general, the regions appeared to be functioning well in the face of the usual challenges in this type of endeavor. Staff morale and loyalty to Reclamation's mission are commendable. Nevertheless, Reclamation, like most federal agencies, is challenged by changing requirements and the need to maintain its core competencies.

Each of the five regions is responsible for sustaining a significant portfolio of facilities. The committee saw examples of excellence; however, in general, the regions will need to evaluate their asset inventory and manage their assets more aggressively and engage in constructive relationships with customers and stakeholders. If Reclamation wants to demonstrate consistency throughout the organization under its style of decentralized management, it will need clear, detailed policy directives and standards to enable all elements to implement a uniform, structured approach. A delicate balance needs to be maintained so as not to impede decentralized units from demonstrating initiative and increasing their capabilities. At the same time, the committee emphasizes that the bureau as the owner has the responsibility to ensure that its facilities are planned, designed, constructed, and managed with a level of quality that is consistent throughout the organization.

The committee believes that Reclamation will continue to have a need for centralized technical services, research, and oversight to support the local management of resources but also sees a need to evaluate the size and organization of the central units to ensure that services are delivered efficiently and at a reasonable cost to Reclamation customers. Both the organization and quantity of services provided at the central, regional, and area offices are affected by the current practice of outsourcing services for constructing, operating, and maintaining facilities and infrastructure that are not inherent to the government's roles and responsibilities.

The committee recognizes that organizations can and do take on a variety of forms with varying degrees of success. Some will function successfully despite their form, while others will falter even as they deploy the best of theoretical forms. The internal culture and history of an organization play a significant role in determining the appropriate structure and the ultimate outcome. The committee believes that the organization of Reclamation is appropriate for its customer-driven mission to deliver power and water. The committee also believes that there are opportunities for Reclamation to improve the construction and management of its facilities and infrastructure and the management, development, and protection of water and related resources in an environmentally sound manner in the interest of the American public. These opportunities are described in the following findings and recommendations.

FINDINGS AND RECOMMENDATIONS

Centralized Policy and Decentralized Operations

Finding 1a. For the past decade many of Reclamation's functions have been decentralized and directed by regional office directors and area office managers. Concurrent with implementation of the decentralized organizational model, Reclamation-wide directives, known as Instructions, were withdrawn, although in some cases they continue to be used for guidance in the field. Mandatory requirements that replace the Instructions have been and continue to be developed and published as policy

---

[1]The *Reclamation Manual* is a Web-based collection of policies and directives that is continually updated and revised. Available at http://www.usbr.gov/recman/.

EXECUTIVE SUMMARY

and directives in the *Reclamation Manual*.[1] However, some issues either have not been addressed or need additional detail. This has led to inconsistencies in understanding and implementing the functions to be performed at each level of the organization, the standards to be applied, and the authority and accountability at each level. Consistently implementing Reclamation's mission will require clear statements of policy and definitions of authority and standards.

Finding 1b. Reclamation's customers and other stakeholders want close contact with empowered Reclamation officials. They also want consistency in Reclamation policies and decisions and decision makers with demonstrated professional competence.

Finding 1c. Decentralization has meant that some area and project offices housing a dedicated technical office are staffed by only one or two individuals. The committee is concerned about the effectiveness of such small units and whether their technical competencies can be maintained.

Recommendation 1a. To optimize the benefits of decentralization, Reclamation should promulgate policy guidance, directives, standards, and how-to documents that are consistent with the current workload. The commissioner should expedite the preparation of such documents, their distribution, and instructions for their consistent implementation.

Recommendation 1b. Reclamation's operations should remain decentralized and guided and restrained by policy but empowered at each level by authority commensurate with assigned responsibility to respond to customer and stakeholder needs. Policies, procedures, and standards should be developed centrally and implemented locally.

Recommendation 1c. The design groups in area and project offices should be consolidated in regional offices or regional technical groups to provide a critical mass that will allow optimizing technical competencies and providing efficient service. Technical skills in the area offices should focus on data collection, facility inspection and evaluation, and routine operations and maintenance (O&M).

Technical Service Center

Finding 2a. The Technical Service Center (TSC) is a large, centrally located, highly structured organization with numerous separate subunits. Many Reclamation customers and stakeholders believe that its costs are excessive, it imposes overly stringent requirements, it too often fails to complete specified work on time, and it sometimes executes projects in a manner contrary to the concept of decentralization. The size of TSC is perceived to be excessive and its organization to be inefficient.

Finding 2b. TSC's response to criticisms has been to benchmark itself against private sector architecture and engineering organizations and to adopt some private sector business practices. In an effort to remain cost competitive, TSC has developed a business plan that provides some services that are not inherently governmental.[2] A strategy of cost averaging, which blends the costs of specialized technical services and oversight with those of other services such as collection of field data and development of construction documents, will continue to subject TSC to fire from Reclamation customers and its private sector competitors and is inconsistent with current federal outsourcing initiatives.

Finding 2c. Regional offices, area offices, water and power beneficiaries, and other stakeholders all perceive an ongoing need for a centralized, high-level center of science and engineering excellence within Reclamation. The committee believes that a thorough review and evaluation of TSC and its policies and procedures could result in a smaller, more efficient and effective TSC.

Recommendation 2a. The commissioner should undertake an in-depth review and analysis of TSC to identify the needed core technical competencies, the number of technical personnel, and how TSC should be structured for maximum efficiency to support the high-level and complex technical needs of Reclamation and its customers. The proper size and composition of TSC are dependent on multiple factors, some interrelated:

- Forecast workload,
- Type of work anticipated,
- Definition of activities deemed to be inherently governmental,
- Situations where outsourcing may not be practical,
- Particular expertise needed to fulfill the government's oversight and liability roles,

---

[2]The basic definition of an inherently governmental function from Office of Management and Budget Policy Letter 92-1 is as follows: "As a matter of policy, an 'inherently governmental function' is a function that is so intimately related to the public interest as to mandate performance by Government employees. These functions include those activities that require either the exercise of discretion in applying Government authority or the making of value judgments in making decisions for the Government." See Chapter 3 for a detailed discussion.

EXECUTIVE SUMMARY

- Personnel turnover factors that could affect the retention of expertise, and
- The need to maintain institutional capability.

This assessment and analysis should be undertaken by Reclamation's management and reviewed by an independent panel of experts, including stakeholders.

Recommendation 2b. The workforce should be sized to maintain the critical core competencies and technical leadership but to increase outsourcing of much of the engineering and laboratory testing work.

Recommendation 2c. Alternative means should be developed for funding the staff and operating costs necessary for maintaining core TSC competencies, thereby reducing the proportion of engineering service costs chargeable to customers.

Reclamation Laboratory and Research Activities

Finding 3. Reclamation's laboratory and research activities came of age during the era of large dam construction in the twentieth century, when much of the needed expertise resided in the federal government and there were no laboratories capable of handling the necessary work. The needs for large materials, hydraulics, and geotechnical laboratories are much different today because the types of capabilities needed to carry out Reclamation's mission have evolved and are available from other organizations (government, university, and private). Although the need for research on environment and resource management continues to grow, the committee believes that the laboratory organization and its physical structure may be too large.

Recommendation 3a. Reclamation's Research Office and TSC laboratory facilities should be analyzed from the standpoint of which specific research and testing capabilities are required now and anticipated for the future; which of them can be found in other government organizations, academic institutions, or the private sector; which physical components should be retained; and which kinds of staffing are necessary. The assessment should also recognize that too much reliance on outside organizations can deplete an effective engineering capability that, once lost, is not likely to be regained. In making this assessment Reclamation should take into account duplication of facilities at other government agencies, opportunities for collaboration, and the possibility for broader application of numerical modeling of complex problems and systems.

Recommendation 3b. Considering that many of the same factors that influence the optimum size and configuration of the TSC engineering services also apply to the research activities and laboratories, Reclamation should consider coordinating the reviews of these two functions.

Outsourcing

Finding 4a. From its inception, Reclamation has undertaken difficult, highly technical projects with a talented and dedicated workforce of engineers and craftsmen. Reclamation's tasks have changed and the composition of its workforce has changed accordingly, but it continues to be an organization that primarily executes engineering and construction for O&M and some rehabilitation and modernization. Reclamation has been outsourcing some of its O&M functions, primarily in nontechnical areas, but could outsource more. The committee believes that many of Reclamation's activities are not what would generally be considered essentially governmental. The committee further believes that although water operations policy decisions are essentially governmental, implementation of these decisions is not and could be almost completely outsourced.

Finding 4b. Decisions on which personnel to use—area, regional, TSC, or contractors—tend to be made at the regional level and on an ad hoc basis. Decisions often hinge on the availability of federal employees to do the work. There is increasing pressure on Reclamation to allow water districts, American Indian tribes, and other customers to undertake their own planning, design, and construction management functions.

Recommendation 4. Reclamation should establish an agency-wide policy on the appropriate types and proportions of work to be outsourced to the private sector. O&M and other functions at Reclamation-owned facilities, including field data collection, drilling operations, routine engineering, and environmental studies, should be more aggressively outsourced where objectively determined to be feasible and economically beneficial.

Planning for Asset Sustainment

Finding 5a. The committee observed effective systems for planning and executing facility O&M in some regions. The 5- and 10-year plans based on conditions assessments and maintenance regimes form the core of the process. The result is an infrastructure that appears able to support Reclamation's mission for the foreseeable future.

EXECUTIVE SUMMARY 9

Finding 5b. The O&M burden for an aging infrastructure will increase, and the financial resources available to Reclamation, its customers, and contractors may not be able to keep up with the increased demand. Some water customers already find full payment of O&M activities difficult, and major repairs and modernization needs, if included in the O&M budget, impose an even greater financial burden that cannot be met under the current repayment requirements. Long-term sustainment will require more innovation and greater efficiency in order to get the job done.

Finding 5c. The committee observed extensive efforts and success in benchmarking Reclamation's hydropower activities; however, there appears to be little effort to benchmark the O&M of water distribution facilities. The committee believes that benchmarking can help improve the efficiency of Reclamation's water management and distribution activities as well as those of the water contractors responsible for transferred works.

Recommendation 5a. Because effective planning is the key to effective operations and maintenance, Reclamation should identify, adapt, and adopt good practices for inspections and O&M plan development for bureauwide use. Those now in use by the Lower Colorado and Pacific Northwest regions would be good models.

Recommendation 5b. Reclamation should formulate comprehensive O&M plans as the basis for financial management and the development of fair and affordable repayment schedules. Reclamation should assist its customers in their efforts to address economic constraints by adopting repayment procedures that ease borrowing requirements and extend repayment periods.

Recommendation 5c. Benchmarking of water distribution and irrigation activities by Reclamation and its contractors should be a regular part of their ongoing activities.

Project Management

Finding 6a. Reclamation does not have a structured project management process to administer planning, design, and construction activities from inception through completion of construction and the beginning of O&M. Projects are developed in three phases: (1) planning (including appraisal, feasibility, and preliminary design studies), (2) construction (including final design), and (3) O&M, with each phase having a different management process.

Finding 6b. The *Reclamation Manual* includes a set of directives for managing projects, but it is incomplete, and there is insufficient oversight of its implementation. Central oversight of some projects is being developed in the Design, Estimating, and Construction Office, but policies and procedures have not yet been completed.

Finding 6c. Reclamation needs to recognize project management as a discipline requiring specific knowledge, skills, and abilities and to require project management training and certification for its personnel who are responsible for project performance. The committee observed the appointment of activity managers in the Pacific Northwest region who were responsible for communications and coordination among project participants for all phases of the project. These activity managers appeared to be beneficial for the execution of projects, but the committee believes that a project manager with responsibility and authority to oversee projects from inception to completion could be even more effective.

Finding 6d. Reclamation has long-standing experience and expertise in planning, designing, and constructing water management and hydroelectric facilities, yet recurring problems are affecting the agency's credibility for estimating project costs. The cost estimating problems associated with the Animas–La Plata Project are a notable example. This project was submitted for appropriations with an incomplete estimate and became a serious problem for Reclamation. Comprehensive directives on the cost estimating process have been drafted but have not yet been published. These directives require that a feasibility estimate be completed before a project is submitted for appropriations.

Recommendation 6a. Reclamation should establish a comprehensive and structured project management process for managing projects and stakeholder engagement from inception through completion and the beginning of O&M.

Recommendation 6b. Reclamation should develop a comprehensive set of directives on project management and stakeholder engagement that is similar to TSC directives for agency-wide use.

Recommendation 6c. Reclamation should establish a structured project review process to ensure effective oversight from inception through completion of construction and the beginning of O&M. The level of review should be consistent with the cost and inherent risk of the project and include the direct participation of the commissioner or his or her designated representative in oversight of large or high-risk projects. The cri-

EXECUTIVE SUMMARY

teria for review procedures, processes, documentation, and expectations at each phase of the project need to be developed and applied to all projects, including those approved at the regional level.

Recommendation 6d. A training program that incorporates current project management and stakeholder engagement tools should be developed and required for all personnel with project management responsibilities. In addition, project managers should have professional certification and experience commensurate with their responsibilities.

Recommendation 6e. Reclamation should give high priority to completing and publishing cost estimating directives and resist pressures to submit projects to Congress with incomplete project planning. Cost estimates that are submitted should be supported by documents for design concept and planning, environmental assessment, and design development that are sufficiently complete to support the estimates. Reclamation should develop a consistent process for evaluating project planning and the accuracy of cost estimates.

Acquisition and Contracting

Finding 7. Different Reclamation regions employ different contracting approaches and use a variety of contracting vehicles to meet their acquisition needs. These range from indefinite delivery/indefinite quantity (IDIQ) contracts with multiple vendors to reverse auction or performance-based contracting techniques to achieve more cost-effective results. In addition, some regions are employing innovative approaches for maintaining stakeholder involvement in the contracting process.

Recommendation 7. Reclamation should establish a procedure and a central repository for examples of contracting approaches and templates that could be applied to the wide array of contracts in use. This repository should be continually maintained and upgraded to allow staff to access lessons learned from use of these instruments.

Relationships with Sponsors and Stakeholders

Finding 8. The committee believes that the key to effective relationships between Reclamation and its sponsors and stakeholders is open communication and an inclusive process for developing measures of success. In addition, the more transparent and consistent the processes used by Reclamation, the easier it will be to obtain buy-in from sponsors and stakeholders. The Lower Colorado Dams Office's interactions with its coordinating committee of sponsors illustrate the beneficial effects of these factors and their contribution to successful operation of the project.

Recommendation 8. Making information readily available about processes and practices, both in general and for specific projects and activities, should be a Reclamation priority. Successful practices, such as those used in the Lower Colorado Dams Office, should be analyzed and the lessons learned should be transferred, where practical, throughout the bureau.

Workforce and Human Resources

Finding 9a. Reclamation and other federal agencies recognize that successful outsourcing of technical services requires maintaining technical core competencies to develop contract scope, select contractors, and manage contracts and that it is necessary for agency personnel to execute projects as well as to receive continuing training in order to maintain those competencies.

Finding 9b. Reclamation's current work is dominated by two categories of tasks: (1) the operation, maintenance, and rehabilitation of existing structures and systems and (2) the creation and brokering of agreements among a variety of groups and interests affected by the management of water resources. The need to include a broad spectrum of stakeholders, particularly groups that represent environmental issues and American Indian water rights, considerably alters both the tasks of the bureau and the skills required to accomplish them.

Finding 9c. Reclamation employees appear on the whole to be more motivated by complex technical tasks than by tasks that are socially and politically complex. However, an increasing proportion of the work that employees at all levels engage in involves tasks that are socially and politically complex. Reclamation's current mission requires personnel to be equipped to address both technical uncertainties and the ambiguities of future social and environmental outcomes.

Recommendation 9a. Reclamation should do an analysis of the competencies required for its personnel to oversee and provide contract administration for outsourced activities. Training programs should ensure that those undertaking the functions of the contracting officer's technical representative are equipped to provide the appropriate oversight to ensure that Reclamation needs continue to be met as mission execution is transferred.

Recommendation 9b. In light of the large number of retirements projected over the next few years and the potential loss of institutional memory inherent in these retirements, a formal review should be conducted to determine what level of core capability should be maintained to ensure that Reclamation remains an effective and informed buyer of contracted services.

Recommendation 9c. Reclamation should recruit, train, and nurture personnel who have the skills needed to manage processes involving technical capabilities as well as communications and collaborative processes. Collaborative competencies should be systematically related to job categories and the processes of hiring, training, evaluating the performance of, and promoting employees.

Recommendation 9d. Reclamation should facilitate development of the skills needed for succeeding at socially and politically complex tasks by adapting and adopting a small-wins[3] approach to organizing employee efforts and taking advantage of the opportunities to celebrate and build on successes.

Alternative Scenarios for Future Infrastructure Management

Finding 10. While the committee recognizes that the major changes suggested by the alternative scenarios are inappropriate for immediate implementation, the continuation and intensification of identified trends, as described in this report, could lead to a need for dramatic changes in Reclamation's operations and procedures in the years to come. The three future scenarios presented in this report—(1) a centrally located project management organization, (2) outsourced O&M, and (3) federal funding and local execution—provide a basis for anticipating future trends and preparing for future change.

---

[3]A small-wins strategy involves working on complex social problems by laying out tasks that can be accomplished without a huge amount of coordination. This strategy puts more control in the hands of individuals, reduces anxiety levels, and makes it possible for people to succeed in ways that can be celebrated and built upon. (See Chapter 4, section on employee motivation, for a description of this strategy.)

Recommendation 10. Reclamation should consider the suggested future scenarios as a basis for analyzing longer-term trends and change.

REFERENCE

U.S. Bureau of Reclamation (USBR). 2005. "Bureau of Reclamation—about us." Available at http://www.usbr.gov/main/about/. Accessed July 29, 2005.

1

Introduction

BACKGROUND

In the more than 100 years since President Theodore Roosevelt signed the Reclamation Act (U.S. Congress, 1902), the U.S. Bureau of Reclamation (Reclamation) has compiled an enviable record, and it can take justifiable pride in having brought water and electrical power to the arid regions of the 17 western states. The dams, reservoirs, hydroelectric plants, and massive irrigation systems developed by Reclamation have been crucial for the development of agriculture and, more recently, for industrial, commercial, and residential development that would not otherwise have been possible. Reclamation is the largest water wholesaler in the country, providing 10 trillion gallons of water to more than 31 million people and irrigating 10 million acres that produce 60 percent of the nation's vegetables and 25 percent of its nuts and fruits (USBR, 2005). It is the nation's second largest producer of hydroelectric power: 42 billion kilowatt-hours of electricity annually. It also partners in the management of more than 300 recreation sites.

This impressive record of accomplishment has been achieved as a result of (or, sometimes, in spite of) complex and overlapping authorizing legislation, regulations, and political pressures; competing local and regional interests; budgetary constraints; and changing national priorities.

SUMMARY OF AUTHORIZING LEGISLATION

The Reclamation program was established by the Reclamation Act of June 17, 1902. The Reclamation Act provided for contracts, generally 10 years in duration, between the United States and individual landowners. Irrigation was the only authorized purpose for a project, and there was a limit of 160 acres per individual. The Reclamation Act required the secretary of the interior to proceed in conformance with state laws as they related to the control, appropriation, use, or distribution of water; however, title to Reclamation projects was to remain with the United States until otherwise provided by the Congress (USBR, 1972).

In 1911 the Congress passed the Warren Act, which authorized Reclamation to contract for conveyance and storage of nonproject irrigation water in project facilities. In 1992 it expanded this authorization to include the conveyance and storage of nonproject water for domestic, municipal, fish and wildlife, industrial, and other beneficial purposes for facilities associated with several non-Reclamation projects in California and Nevada (the Central Valley Project, the Cachuma Project, the Truckee Storage Project, and the Washoe Project) (USBR, 2001).

The Reclamation Extension Act, passed in 1914, provided for the extension of individual repayment contracts for up to 20 years. It also required the payment of operating and maintenance costs and recognized legally organized water users' associations and irrigation districts. Furthermore, it authorized the transfer of project facilities operations and maintenance (O&M) to water districts (USBR, 1972).

In 1920 Congress passed legislation entitled Sale of Water for Miscellaneous Purposes. For the first time, Reclamation was provided authority to contract for the purchase of water for uses other than irrigation. However, such contracts required (1) that no other practicable source of water be available, (2) a finding that such contracts would not be detrimental to the quantity and quality of irrigation water from the project, and (3) the approval of the relevant water users' association.

The Irrigation Districts and Farm Loans Act of May 15, 1922, required that a court of competent jurisdiction confirm contracts between the secretary of the interior and irrigation districts to ensure that the districts had the necessary authority before the contracts became binding. This requirement was reiterated in the Omnibus Adjustment Act of 1926, which also provided that no water could be delivered until a contract was executed, extended the maximum repayment period to 40 years, and established the requirement that O&M costs be paid in advance (USBR, 1972).

On August 4, 1939, Congress passed the Reclamation Project Act of 1939. This act made several significant changes and additions to Reclamation's contracting authority. It provided authority for project costs to be allocated between reimbursable and nonreimbursable purposes, authorized a ceiling on charges to irrigators based on an ability-to-pay concept, and provided authority for the secretary to defer repayment obligations under certain circumstances. It also provided for reimbursable project costs associated with irrigation or municipal and industrial purposes to be recovered through either repayment or water service contracts (USBR, 1972).

The 1956 Act (Administration of Contracts under Section 9, Reclamation Project Act of 1939) assured that contracts would be renewed upon expiration, assured water users they would be relieved of payment for construction charges after the United States had recovered its entire irrigation investment, and assured water users of a first right to contract for the use of water under water-service-type contracts (USBR, 1989).

MISSION

Reclamation started with a focus on irrigation in 16 western states in 1902 (Texas was added in 1906) and quickly (by 1906) evolved into an organization with a core mission to develop and deliver water and hydroelectric power in the West. The current mission statement for the Bureau of Reclamation is as follows: "The mission of the Bureau of Reclamation is to manage, develop, and protect water and related resources in an environmentally and economically sound manner in the interest of the American public" (USBR, 2005).

Carrying out the core mission in the early years of the twenty-first century is much different than it was during most of the twentieth century. Before the 1970s, developing and delivering water and power was dominated by the construction of large dams—for example, Hoover, Grand Coulee, and Glen Canyon (see Figure 1-1)—power plants, and irrigation systems. Reclamation has been responsible for numerous pioneering and world-class engineering and construction accomplishments. Now the focus has shifted. Most large reservoir and hydroelectric sites have been developed. The predominant workload has changed from new construction to the O&M, repair, and modernization of aging infrastructure, evaluation of dam safety and mitigation of dam failure risk, and environmental restoration and enhancement. Water rights issues, pressure from water and power user groups, cost recovery considerations, facility title transfer agreements, and environmental regulations (such as the National Environmental Policy Act, which requires detailed environmental impact studies and statements, and the Endangered Species Act)—all have had major impacts on what Reclamation does and how it does it, and there is no reason to believe that these factors will not be even more important in the years ahead.

FIGURE 1-1 Reclamation's flagship facilities: (upper left) Grand Coulee Dam, (right) Hoover Dam, (lower left) Glen Canyon Dam. SOURCE: USBR.

STATEMENT OF TASK

In response to a request from the Department of the Interior's assistant secretary for water and science, the NRC was asked to form a committee under the Board on Infrastructure and the Constructed Environment to advise the department and the Bureau of Reclamation on the appropriate organizational, managerial, and resource configurations to meet Reclamation's construction, maintenance, and infrastructure requirements for its missions of the twenty-first century. A committee familiar with ongoing changes in the federal civil service system and with alternative means of ensuring organizational core competencies was drawn from industry, academia, and government. Committee members have experience and expertise in water resources facilities engineering, infrastructure management, project delivery methods, federal contracting practices, business process reengineering, and human resources. See Appendix A for biographies of the committee members.

INTRODUCTION

The committee was assigned the following specific tasks:

- Examine the requirements of the Bureau of Reclamation regarding construction, heavy maintenance, and infrastructure operations.
- Survey federal agencies and other governmental and nongovernmental organizations with similar mission responsibilities to determine their organizational and operating models and to identify good practice tools and techniques for Reclamation's efforts in infrastructure management.
- Review and assess trends in budget, human resources, and project execution methods at Reclamation.
- Construct alternative scenarios for future infrastructure management responsibilities and develop corresponding organizational options.

To accomplish these tasks, the committee met as a whole four times from February to August 2005, and small groups visited offices and projects in each of the five Reclamation regions: Great Plains, Upper Colorado, Lower Colorado, Mid-Pacific, and Pacific Northwest. The committee received briefings from and discussed all major activities related to facilities and infrastructure with Reclamation representatives in Washington, D.C., and at Policy, Management, and Technical Services in Denver, Colorado. The committee also met with some of Reclamation's water and hydroelectric customers, organizations representing customer interests, environmental advocates, other federal and state agencies with similar missions, and congressional staff concerned with water issues.

In addition to the knowledge it gained from the references listed in the report, the committee learned from the five regional offices' written responses to 33 questions, meant to provide background information on the organization and activities in their respective regions. Discussion questions were used to guide informal dialogue between Reclamation personnel and committee members during their site visits. Similar questions were also used to guide discussions with Reclamation customers and contractors. To promote open and candid discussion, participants were assured that comments would not be attributed to specific individuals. After completing all of the site visits, the groups reported and discussed their findings with the full committee. The committee's meeting and site visits are listed in Appendix B of this report, along with the questions used to elicit background information.

ORGANIZATION OF THE REPORT

This report is organized first into chapters that present the committee's observations and responses to the four parts of the statement of task. These chapters are followed by a chapter containing the committee's conclusions, findings, and recommendations. Biographies of committee members, a list of meetings and briefings, and a detailed summary of a roundtable discussion with other organizations having similar or related water resources missions are contained in Appendixes A, B, and C.

Chapter 2, "Requirements for the 21st Century," describes the facilities and infrastructure requirements of Reclamation and the factors that will influence future changes in these requirements. Requirements are addressed in terms of the bureau's mission, its management of assets, and other factors that define the work Reclamation needs to accomplish. The policies, procedures, decision-making processes, and organizational structure needed to optimize Reclamation's capabilities are discussed using the 1993 *Blueprint for Reform* as the baseline (USBR, 1993).

Chapter 3, "Good Practice Tools and Techniques," draws on the committee's experience and expertise, discussions with organizations having missions similar to that of Reclamation, and observations gained from discussions with Reclamation personnel and its customers and stakeholders. Tools and techniques for developing policies and procedures, acquisition and contracting, project management, asset management, and planning and budgeting are described. The chapter also reports on general observations from a roundtable discussion with representatives of the U.S. Army Corps of Engineers, the Tennessee Valley Authority, and the California Department of Water Resources.

Chapter 4, "Workforce and Human Resources," discusses strategies for workforce planning to meet the uncertainties and ambiguities that will challenge Reclamation personnel in the future. Following the outline of Reclamation's *Workforce Plan FY 2004-2008,* the chapter assesses strategic direction, supply of and demand for human resources, deficiencies and strategies for mitigating them, and approaches to measuring the bureau's performance in workforce management.

Chapter 5, "Alternative Scenarios for Future Infrastructure Management," presents three scenarios that are considered by the committee to describe possible futures for Reclamation: (1) a centrally located project management organization, (2) outsourced operations and maintenance, and (3) federal funding and local execution. The chapter concludes with a discussion of how Reclamation can use this information to begin planning for the future.

Chapter 6, "Conclusions, Findings, and Recommendations," is based on the discussion in Chapters 2 through 5. It describes the factors affecting the management of construction and infrastructure and the capabilities that will be needed to successfully respond to their impacts. Findings and recommendations are presented for policy development and organization, the Technical Service Center, the research program, outsourcing, asset sustainment, project management, acquisition and contracting, relationships customers and stakeholders, workforce and human resources, and future scenarios.

The report includes three appendixes, "Biographies of Committee Members," "Briefings to the Committee and Discussions," and "Good Practice Tools and Techniques Roundtable."

REFERENCES

U.S. Bureau of Reclamation (USBR). 1972. *Federal Reclamation and Related Laws Annotated, Volumes I-III*. Washington, D.C.: U.S. Bureau of Reclamation.

USBR. 1989. *Federal Reclamation and Related Laws Annotated, Volume IV, Supplement I*. Washington, D.C.: U.S. Bureau of Reclamation.

USBR. 1993. *Blueprint for Reform*. Washington, D.C.: Department of the Interior.

USBR. 2001. *Federal Reclamation and Related Laws Annotated, Volume V, Supplement II*. Washington, D.C.: U.S. Bureau of Reclamation.

USBR. 2005. "Bureau of Reclamation—about us." Available at http://www.usbr.gov/main/about/. Accessed July 29, 2005.

U.S. Congress. 1902. *Reclamation Act/Newlands Act of 1902*, P.L. 161, 57th Cong. 1st Sess., CH. 1093, June 17, 1902.

2

Requirements for the Twenty-first Century

INTRODUCTION

Reclamation's facility and infrastructure requirements derive from its mission. The bureau presents its mission in two ways. The first, "Delivering water and generating power, and whatever it takes to do these," was relayed to the committee at briefings and meetings. The second, as posted on Reclamation's Web site is this: "The Bureau of Reclamation manages, develops and protects water and related resources in an environmentally and economically sound manner in the interest of the American public" (USBR, 2005a). The first characterization focuses on the bureau's output and seems to be oriented to breaking through the barriers to delivering water and generating power. This statement of mission would have been applicable in the twentieth century, when the barriers were mountains and river valleys and the problem was how to build big dams that were safe, effective, and efficient. The second version recognizes the twenty-first century tasks and processes that the bureau needs to engage in to accomplish its desired outcomes, which are quite different than they were in Reclamation's earlier years. Delivering water and power today includes negotiating American Indian water rights, working with environmental groups to agree on reasonable ways to protect the environment and endangered species, and finding ways to promote water conservation. The second mission statement better portrays what Reclamation actually does.

A Web-based orientation to the Department of the Interior presents Reclamation's evolving mission as the following (DOI, 2005):

Reclamation's evolving mission places greater emphasis on water conservation, recycling, and reuse; developing partnerships with our customers, states, and tribes; finding ways to bring competing interests together to address everyone's needs; transferring title and operation of some facilities to local beneficiaries who might more efficiently operate them and achieving a higher level of responsibility to the taxpayer.

This statement does not, however, elaborate the role that Reclamation plays in water conservation, developing partnerships, managing assets, and so forth. As the statement suggests, the role is evolving, and changes in asset management processes, workload, and organization will be needed.

In the twentieth century Reclamation's goals were about developing facilities and infrastructure and the resources to foster development of the West. Today, its goals are about sustaining its facilities, infrastructure, and resources, as well as responsibly managing the environment. This shift was addressed to some degree in the bureau's reorganizations in the 1980s and 1990s, and Reclamation continues to adapt to evolving goals and shifting obstacles.

FACILITY AND INFRASTRUCTURE ASSETS

Ownership of Assets

Since the creation of the Bureau of Reclamation in 1902, the organization has designed and constructed a wide variety of physical facilities to manage water resources and generate electric power in 17 western states. Reclamation's inventory of facilities and infrastructure is large and diverse in both size and type. The inventory is the result of the water and power projects that have been authorized by Congress. Using the number of projects as a measure can be somewhat misleading, because they vary in size and complexity from a single canal distribution system, such as the Avondale Project, near Hayden Lake, Idaho, to large, complex, multifeature projects, such as the Colorado–Big Thompson (CBT) in Colorado, which consists of 17 facilities, including dams, hydroelectric plants, canals, tunnels, and pumping plants. One feature of the CBT, the Horsetooth Dams, is considered to be a single facility but consists of four dams and a dike.

Depending on definitions and counting procedures, Reclamation's inventory includes about 673 facilities that have been constructed as part of 178 major projects. Included in this inventory are 471 dams and dikes, 58 hydroelectric plants, and more than 300 associated features such as canal systems, pumping plants, pipeline systems, fish protection facilities, diversion and drainage facilities, structures, and buildings (Keys,

2005; USBR, 2000, 2005a). Although difficult to count, the number of facilities currently owned by Reclamation appears to be relatively stable, requiring an effective management strategy and a focus on operations, maintenance, repair, and modernization rather than development.

Reclamation's objective is to transfer ownership of as many noncritical or low-risk assets as possible to the beneficiaries of the resources. Since 1995, Reclamation has transferred title to 18 projects and parts of projects, and it is finalizing the transfer of 5 more that were authorized for transfer by Congress. Of the 18, four were entire projects (Middle Loup in Nebraska, Palmetto Bend in Texas, and Sly Park and Sugar Pine in California) and the rest were distribution facilities and associated lands. However, it appears that very few additional assets currently owned by the bureau will ever be transferred. The issues of dam safety, security, and reliability of power generation make it difficult to transfer the hydroelectric facilities or the other large dams. In addition, the costs associated with operations and maintenance (O&M) are prohibitive for small irrigation districts, and it is expected that they will continue to resist incurring the responsibilities, liabilities, and costs that would be associated with ownership. Even large, self-sustaining districts like the Central Utah Project see a benefit in continued federal ownership of the facilities. Therefore, unless funding mechanisms are changed, Reclamation will continue to be responsible for many facilities and a large infrastructure for the foreseeable future.

Management of Assets

Reclamation's assets are managed by 24 area offices organized on a regional basis, with each of the five regional offices having full responsibility for operating and maintaining the assets in its region. In most cases this means that all the assets in a single watershed are operated and maintained by the same regional office. However, two regional offices are responsible for the operation of the facilities in some basins, such as the Colorado River, Canadian River, and Rio Grande River basins, necessitating an additional level of coordination.

The committee discussed the advantages and disadvantages of watershed management and project management. Because Reclamation is one of many organizations, including the U.S. Army Corps of Engineers (USACE) and state agencies, that have decision-making authority for water use and distribution in the watersheds where Reclamation operates, the committee concluded that it would not be possible for Reclamation to manage its assets strictly on a watershed basis. It would probably be more efficient to have the water managed on a basinwide basis, but the current set of water laws and diverse management agendas and stakeholder interests pose challenges for such an approach.

Within the regions, the facilities tend to be managed on a portfolio basis, with each project competing with the others in the region for funding and personnel. The main driver for decision making appears to be the budgeting process. In addition, the bureau also oversees O&M activities at facilities where the O&M responsibilities have been transferred to local beneficiary organizations. The committee discussed the possible benefits of additional transfer of O&M responsibilities to users, with proper oversight by Reclamation. In most of these cases, however, it would be difficult to do so, partly because there is no way for Reclamation to help to build an O&M capacity within the user organizations. Such capacity depends on resources and initiative: Organizations that have the will and resources have generally built the capacity and those that do not continue to rely on Reclamation. However, this does not preclude outsourcing O&M activities.

Adaptive Management of Resources

Demands on water management agencies have increased in complexity, fervor, and emotion, and Reclamation has worked to adapt its management strategies to deal with this changing landscape. As the availability of water stays steady or decreases due to weather patterns in the West and as the demand for water—from existing users as well as new users such as urban systems and environmental enhancement—increases, better methods will be needed for decision making, communication, and engaging stakeholders. Reclamation uses adaptive strategies to satisfy as many of the demands as possible. This approach uses scientific information to improve procedures and enhance fish habitat and survival. Reclamation has also begun to apply these adaptive strategies to mitigation activities not directly associated with Reclamation projects, and the demand for such services is expected to increase (NRC, 2004).

Identification of Needs

Since the assets managed by Reclamation have an average age of more than 50 years and require almost constant review and upgrading, the area and regional offices have ongoing procedures for identifying needs. In his testimony before the House of Representatives Subcommittee on Water and Power, Commissioner Keys noted as follows (Keys, 2005, p. 1):

> Some components and replaceable equipment have well-defined design and service lives while many of the larger structures do not. In many cases the estimated service lives have been and continue to be exceeded.

Reclamation attributes its success in lengthening these service lives to a commitment to preventive maintenance that has guided our O&M practices over the years.

However, there is inconsistency in the way these processes operate and in how the beneficiaries are engaged in decision making and review. Some beneficiaries noted that the rules seem to differ within regions and across regions with respect to who must pay, how much must be paid, and how design and construction activities are carried out. The quality and consistency of assessment and planning documents, except those associated with the larger power facilities, also vary from region to region.

Availability of funding is an important factor in setting priorities. This can create constructive tension in the prioritization process, but when resources are too limited, the process can be distorted. Several regions rely on a bottom-up process from the area offices, driven by the core mission to deliver water and power, using a variety of teams and review processes to finalize priorities on a regional basis. One region reported using a 10-year resource plan as a part of its priority-setting process.

WORKLOAD

Reclamation's facility inventory drives its technical workload, which includes the planning, design, and construction of dams, hydroelectric plants, and related infrastructure. The tasks involve O&M, replacement and modernization, modification to improve dam safety and meet environmental requirements, and new construction. This workload is made more complex by the need to interact with an expanding and increasingly diverse set of stakeholders with growing environmental and social expectations.

Design and Construction of Dams

For 86 years, from the passage of the Reclamation Act in 1902 until 1988, the work of the bureau was dominated by the design and construction of new dams, hydroelectric plants, and irrigation infrastructure (see Figure 2-1).[1] The last large construction authorization was the Colorado River Basin Projects Act in 1968, which also included facilities in the Central Arizona Project, the Central Utah Project, and the Central Valley Project in California. From 1969 through 1988, Reclamation continued to

---

[1]Brit Storey, Reclamation historian, "Organizational history of the Bureau of Reclamation," Presentation to the committee on February 28, 2005.

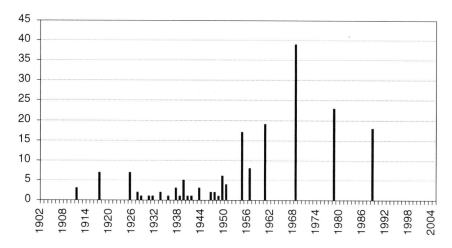

FIGURE 2-1 Reclamation construction projects completed in different years. SOURCE: USBR.

have a heavy construction workload, bringing the total number of completed projects to 178.[2] The year 1988 marked the end of Reclamation's traditional role as a major designer and constructor of new dams. Since 1988, the bureau has continued to design and build new dams, but at a much reduced scale.

Dam Safety

In 1976, the failure of the Teton Dam sparked new interest in dam safety. The Reclamation Safety of Dams Act (P.L. 95-578) was passed in 1978 and amended in 1984. This Act authorized and funded modifications to preserve the structural safety of Reclamation's dams and related facilities. In response to the Teton Dam failure, Reclamation instituted the Safety of Dams Program (SOD) and an extensive safety inspection process for dams determined to pose high and significant hazards (USBR, 1998). The safety evaluation of existing dams (SEED) is the overall process for

---

[2]Follow-up communication with Brit Story indicated that the total of 178 projects is based on several assumptions as to what constitutes a "Reclamation project." For example, it does not include several projects that were not appropriated funds for completion, four dams designed and built for the Bureau of Indian Affairs, the design of one dam at the Panama Canal, and participation in the design of several Tennessee Valley Authority dams. Some projects are consolidations of earlier separate projects.

identifying and evaluating potential risks and determining whether action needs to be taken to reduce risk to the public. The process includes in-depth periodic facility reviews (PFRs) and comprehensive facility reviews (CFRs), which are conducted alternately on 3- and 6-year cycles and supplement the annual O&M inspections. Severe deficiencies and important maintenance needs are tracked through the Dam Safety Information System (DSIS). To date, approximately 3,600 SOD deficiencies have been corrected. Modifications have been made to 69 dams at a cost of $868 million.[3,4]

SOD has become a significant component of the technical workload of the bureau. From FY 1996 through FY 2005, funding for SOD averaged $66 million per year. Many of these projects are as complex as the design of a new dam. In addition, stakeholder and public interest group involvement has increased significantly, much of it concerning environmental issues.

It is realistic to expect that within the foreseeable future, major renovations will be required to address dam safety issues. Currently, 12 additional dams needing modification have been identified, with preliminary cost estimates totaling $350 million. There are also more than 400 incomplete SOD recommendations requiring additional field investigations or engineering analysis to determine if risks are such that action is needed. Most of these recommendations indicate that a dam modification may be necessary to reduce the risk.

Operations and Maintenance of Bureau-Operated Facilities

As the number of completed projects in Reclamation's inventory has risen, so has the O&M workload. Today, O&M is the primary technical workload of the bureau and is likely to remain so because of the aging infrastructure and the need for rehabilitation and modernization of facilities. The average age of completed projects (see Figure 2-1) is approximately 50 years. Some individual facilities are 90 years old. The age of the facilities also means that most embody out-of-date design, engineering practices, and materials. It is estimated that 90 percent of the dams are in this category (Achterberg, 1999).

The maintenance workload and backlog of needs are tracked by a number of methods. In power facilities, Maximo-based computerized

---

[3] Larry Todd, director, Security Safety and Law Enforcement, and Bruce Muller, chief, Dam Safety Office, Briefing to the committee on April 6, 2005.

[4] "Safety of Dams modifications completed," Spreadsheet provided by Dam Safety Office, July 2005.

maintenance management systems are used. Critical maintenance problems receive immediate attention. Less-than-critical needs are prioritized and scheduled as funds become available. At nonpower facilities, needs beyond the scope of normal day-to-day maintenance are tracked through DSIS and replacement, addition, and exceptional maintenance (RAX) lists. The RAX lists are also used to prioritize maintenance needs and funds through the budget formulation process. Budget proposals are generated by the area offices and consolidated at the regional and headquarters levels.

Technical Workload and the Technical Service Center

The technical workload is distributed among the various project, area, and regional offices, and the Technical Service Center (TSC) in Denver. The more routine engineering for O&M and repair are undertaken by the area and regional offices, while TSC provides centralized engineering and scientific services that are typically beyond the capability of the areas and regions.[5] In FY 2004 its workload was distributed among clients approximately as follows: support to regions and areas, 46 percent; safety of dams support, 21 percent; research and development, 7 percent; other Reclamation organizations, 11 percent; and non-Reclamation organizations, 15 percent.

The size and composition of TSC depend on many factors, some interrelated:

- Forecast workload,
- Type of work anticipated,
- Activities deemed to be inherently governmental,
- Areas where outsourcing may not be practical,
- Particular expertise needed to fulfill the government's oversight and liability role,
- Turnover factors that could affect retention of expertise, and
- The need to maintain institutional capability.

At present, TSC employs more than 600 people (down from 800 in 1994) and is funded on a fee-for-service basis. It is essentially a very large service unit without a line-management function. As there is no annual funding for TSC, all salary and overhead costs not directly chargeable to a specific project have to be absorbed by all projects that use TSC services.

---

[5]Michael Roluti, director, Technical Service Center, Briefing to the committee on April 6, 2005.

The committee does not question the need for a technical service unit of this nature within Reclamation, but it does question the size. Reclamation, in its role as an owner, needs to determine which activities performed by TSC are inherently governmental and should not be performed by outsourcing and the quantity and type of engineering that needs to be performed in house in order to maintain the competencies of a smart owner. A smart owner "retains core competences to establish project definitions, establish project metrics, monitor project progress, and ensure commitment, stability, and leadership" (NRC, 2000). By assessing these matters, the bureau can ensure a long-term and stable structure for TSC and its critical support to Reclamation's missions.

A strict reading of Office of Management and Budget (OMB) Circular A-76, *Performance of Commercial Activities* (OMB, 2003), would likely find that only a limited number of the technical activities performed by TSC are inherently governmental functions. (A process for identifying essentially governmental functions is discussed in Chapter 3.) The same would apply to similar activities performed at the regional or area levels. However, other factors warrant consideration. Foremost, it has to be recognized that Reclamation owns a large number of structures and facilities that pose a potential risk to public safety, the national economy, and the environment should one of them fail. As the owner, Reclamation cannot escape liability for any negative consequences if a facility malfunctions no matter who may have designed, constructed, or maintained it. To ensure that these risks are minimized, Reclamation needs to exercise a certain level of oversight and control.

Cost savings are one of the many benefits that might be gained from outsourcing, but they can be the most difficult to assess. Many state governments have found that the cost-benefits of outsourcing are not always clear (Moore, 2000). This might be due to the specialized nature of infrastructure projects, project-to-project variations, and/or the considerable oversight necessary to ensure compliance with agency standards. The cost of oversight and of preparing addenda and change orders to bring engineering designs into compliance with agency standards can cancel out any cost savings realized by using consultants. Design costs are generally lowest when states use a mix of private and public sector work.

Exercising oversight is more than a perfunctory matter and requires particular expertise and knowledge of Reclamation's facilities and infrastructure. Such expertise is derived from the actual performance of the scientific and technical functions inherent in the projects. To develop and maintain the necessary cadre, a certain amount of work must be performed in-house. It would be appropriate for TSC to perform the following kinds of work:

- Development of design standards,
- Design review,
- Cost allocation,
- Cost estimating in the early planning stages,
- Cost estimating for very large or complex facilities,
- Environmental planning, permitting, and mitigation strategy,
- Power plant design and rehabilitation,
- Major dam design and rehabilitation,
- Major pumping plant, tunnel, and canal design and rehabilitation,
- Risk assessment, and
- Project-applicable research.

Determining the optimum size for TSC is a challenge that needs to be addressed by Reclamation. The challenge today is different from that faced after World War II, when major water resource projects were being developed. In that era, much of the expertise in dams and hydraulic structures resided within federal agencies. That is not true today, when private and semipublic organizations have the expertise required to perform many of the functions carried out by Reclamation in the past. The committee foresees the possibility of TSC becoming more involved in oversight and the establishment of standards than in design and construction document development. It appears to the committee that TSC might be able to provide its services for oversight, highly technical design in critical areas, and a limited quantity of design in noncritical areas—and at the same time maintain its core competencies—with a smaller workforce.

Operation and Maintenance of User-Operated Facilities

Some of the facilities and infrastructure inventory are transferred works that are owned by Reclamation but user-operated and -maintained with oversight by Reclamation. Transferred works are generally irrigation-system-related facilities, including smaller dams, dikes, pumping plants, and canals.

The resources and sophistication of the water districts vary. The committee observed that some districts are willing and able to perform a larger role. Some districts feel that they can perform the O&M functions and more complex repair and modernization projects at lower cost than Reclamation by using local staff and contractors.

Future Workload

As Reclamation moves further along the transition announced in 1993 from a water and power construction organization to a water and power management organization, the responsibilities, duties, and activities of the workforce are changing significantly. The workload change is driven by a number of factors, including the following:

- *Aging infrastructure.* Many of the dams and associated conveyance and distribution facilities are over 50 years old, and their maintenance needs are growing as the structures and equipment reach or pass the design lifetimes.
- *Increasing competition for declining resources.* Since water availability continues to decline in many parts of the West, existing water users continue to demand reliable systems to provide the water they have historically used, while new users would like to obtain access to the water or, in some cases, to the land adjacent to the facilities that provide the water services.
- *Increased regulatory requirements.* Water rights regulations, Endangered Species Act (ESA) requirements, environmental impact assessment (EIA) requirements, and expectation for increased openness and public involvement in decision making place additional demands on project managers, operators, and decision makers.
- *Security.* Security reviews and ongoing security management at the existing facilities add to the workload at many of the large facilities operated by Reclamation. Several of these sites are considered national critical infrastructure.

Maintenance activities will grow in complexity and costs as the facilities age, so the depth and breadth of expertise in the areas of project design, cost estimating, and project management will need to be maintained even with increased outsourcing of many activities. However, there will also be an increasing need for expertise in stakeholder engagement, communications, endangered species and environmental requirements, and data collection, as well as for expertise in conducting negotiations among stakeholders with divergent expectations associated with the facilities and the services that Reclamation provides. Reclamation representatives are increasingly expected to take a more active role in the negotiation processes that typically occur when complex water issues are addressed by multiple stakeholders. The complexity of these interactions and the time-consuming processes employed to achieve agreement among the many stakeholders and regulators will burden Reclamation with additional work. Some of the additional responsibilities can be met through the use of outside expertise, but there will remain the need to have enough management and oversight capabilities within Reclamation to ensure that the issues are being addressed properly.

Reclamation has developed some specialized expertise for its internal needs that is also needed by other agencies, state and local governments, and industry. The bureau is moving in the direction of providing services to others for projects not directly related to it own facilities, such as dam removals and environmental mitigation programs. The broader scope of Reclamation's services can be one way of attracting and keeping good employees, but it adds to the agency workload.

MANAGEMENT POLICIES AND PROCEDURES

Prior to 1993, Reclamation had a massive body of policy and procedural directives referred to as the Reclamation Instructions. The Instructions were prescriptive in nature, centralized in origin, and generally reflective of the organizational and management philosophy governing the bureau. However, they contradicted popular management models used by the government in the early 1990s.

Public management reforms widely known as "the new public management" have taken a variety of forms.[6] In New Zealand and Great Britain, what goes by this name emphasizes the proper construction of incentive structures to *"make managers manage"* and is deemed to be the key to government performance. The American version of new public management, by contrast, supports greater management flexibility. Its advocates argue that we should *"let managers manage."* They believe that the work of government would be vastly improved if managers in the public sector had the same flexibility as managers in the private sector so that they would perceive their work in terms of goals such as the "creation of public value" or the pursuit of continuous improvement. Reclamation was one of many federal agencies that adopted these principles of change as part of the 1990s efforts to reinvent government.

Commissioner Beard endorsed the recommendations of a team he had appointed to examine the need for change. He noted that Reclamation was moving forward with the exciting challenges awaiting its water resource managers.[7] The Commissioner's Program and Organization Review Team (CPORT) report was cited as critical in identifying "changes

---

[6]This paragraph is adapted from Feldman and Khademian, 2000, p. 341.

[7]Letter from Commissioner Daniel P. Beard to Reclamation employees dated August 6, 1993.

needed in Reclamation's programs in order to successfully complete the transition from a water resources development agency to a water resources management agency."[8] The commissioner's *Blueprint for Reform* criticized the policy process directives then in use as too detailed and inflexible. He also criticized the required multiple stages of "prior review and approval processes." His plan for reform included the following changes (USBR, 1993b):

- Policy directives will be limited to broad, agency-wide applications that set goals and objectives and establish broad parameters for execution. They will generally require the Commissioner's personal approval before issuance.
- Instructions and standards will intentionally allow responsible line managers an appropriate degree of discretion and judgment in accomplishing their duties.
- Use of the procedures, processes, and methodologies set forth in such manuals and handbooks will not be mandatory.
- In order to ensure that this approach to implementing instructions and technical standards is followed, all existing guidance will be sunset at the end of fiscal year 1995 unless affirmatively retained, or revised and reissued prior to then.

Reclamation undertook substantial—one might even say massive—reorganization and change. Centralized oversight was loosened dramatically as senior management positions were eliminated. Services were centralized for efficiency and economy, but operational authority was delegated downward on the organization chart. The absence of mandatory policy and procedural guidelines resulted in every region developing a unique character.[9] The organization and functions of the regional, area, and project offices began to vary widely. Responding to the Clinton administration's directives for reinventing government, staffing was reduced about 10 percent and 40 project offices were consolidated into 24 area offices.[10]

When committee members visited regional and area offices, they were told that bureau policy decisions lack consistency. User associations such as the National Water Resources Association (NWRA) and the Family Farm Alliance told the committee that Reclamation stakeholders communicate with one another and compare Reclamation policy decisions, and

---

[8]Letter from Commissioner Daniel P. Beard to Reclamation employees dated November 1, 1993, with attached *Blueprint for Reform*.

[9]Robert Johnson, regional director, Lower Colorado Region, "Delivering water and generating power," Briefing to the committee on April 6, 2005.

[10]Brit Storey, Reclamation historian, "Organizational history of the Bureau of Reclamation," Presentation to the committee on February 28, 2005.

from these discussions they concluded that the stakeholders are not being treated equally. The NWRA, in its most recent position paper on the bureau, writes: "However, direct and sudden reversals of program direction and organizational philosophy have had a profoundly negative effect on the organization."[11] When they were invited to recommend constructive changes for the bureau, Reclamation employees from several of the regions spoke of the obvious inconsistency affecting many of the bureau's decisions. The Family Farm Alliance noted an inconsistent Reclamation policy on use of TSC.[12] Other stakeholders reported serious inconsistencies in Reclamation reports on the Animas–La Plata project.[13]

Reclamation leadership appears sensitive to the need to promulgate formal policy directives, and a new manual has been issued (USBR, 2005c). The manual is a Web-based collection of policies and directives that is continually being updated and revised. However, as reported to the committee, this process has been slow and inadequate to date. There is disagreement among stakeholders and Reclamation employees as to just what to do and how far to go in reestablishing published policy documents. Some field personnel admit that they have kept copies of the old Reclamation Instructions, which they routinely, but selectively, use in their area of responsibility. Some Reclamation personnel would welcome reinstatement of Reclamation Instructions in their entirety. Others see the need for selective reinstatement.

DECISION-MAKING PROCEDURES

The scope of the 1993 organizational changes and availability of policies and guidance had a significant effect on the decision-making process within Reclamation. The Reclamation Decision Process Team submitted a report in October 2004 (USBR, 2004). On page 1 the team noted as follows:

> The majority of the decision-making problems they [Reclamation personnel who were surveyed as part of the study] identified were due to unclear roles and responsibilities, the lack of a defined decision-making process, or a combination of both. Interviewees were concerned that failure to acknowledge and correct these problems could result in significant consequences to Reclamation, including loss of agency credibility; increased employee frustrations and a decline in morale; poor account-

---

[11]National Water Resources Association, "Role of the U.S. Bureau of Reclamation in the 21st century," Undated position paper, provided to the committee on June 24, 2005.
[12]Family Farm Alliance, Letter to the committee dated June 18, 2005, with an attached compilation of nine case studies.
[13]Committee member telephone interview with Navajo Nation representatives, June 16, 2005.

ability for decisions and implementation; inefficient use of time, personnel, and financial resources; and loss of control of the decision to others (e.g., Congress, courts, etc.).

The team found that the abandonment of formal decision and planning processes and decentralization of the organizational structure has had a mixed impact. It noted that the best managers profited from the flexibility offered by the new organization; others, however, experienced procedural problems and were challenged by the absence of a formal structure and decision processes.

Reclamation personnel interviewed by the committee generally rated the bureau as having a high level of technical skills, but they were more critical of the bureau's managerial abilities. Some thought that a more focused assignment of responsibility—that is, a shift away from the decision-by-committee approach—is needed. The decentralized organization and the absence of coherent, comprehensive centralized policy and procedures has led to divergent decisions and the complaint by user groups about inconsistency.

The committee is concerned that Reclamation's decentralized and collaborative decision process seems to be missing a clear assignment of responsibility, which is essential for effective decision making. It appears especially elusive when more than one Reclamation element is involved, such as TSC, a regional office, and an area office. Thus, the committee commends Reclamation for taking steps to analyze the decision-making process and develop constructive measures that should improve performance.

ORGANIZATIONAL CONFIGURATION

This section addresses the organizational structure employed by Reclamation to construct and maintain its facilities and infrastructure and execute its mission, as well as what this structure may be in the near future. Reclamation is organized to undertake the following facility and infrastructure functions:

- Managing and maintaining existing assets.
- Ensuring dam safety.
- Planning and developing projects to meet future resource needs.
- Developing alternative means of supplying water.
- Managing a program to enhance water conservation.
- Designing and constructing authorized projects.
- Implementing a water and hydroelectric engineering research program.

- Providing environmental benefits through conservation and environmental remediation and enhancement.

This review of Reclamation's organization is developed with the bureau's changing goals and work requirements in mind.

Present Organization for Managing Facilities and Infrastructure

In the present Reclamation organization most of the activity pertaining to water and power management is centered in two directorates under the commissioner: the Directorate of Policy, Management, and Technical Services (PMTS), which functions primarily as a staff service element, and the Directorate of Operations, which functions as a line-management element. Another unit with facilities and infrastructure functions is the Dam Safety Office, a line-type element under the Directorate for Security, Safety, and Law Enforcement.[14]

The lines of authority for construction projects in Reclamation are somewhat unclear because projects are not structured under a single project manager or integrated project team from inception through completion. Management responsibilities shift as the project progresses through various phases, in part because of the way federal civil works projects are planned and authorized. This has the effect of diffusing responsibility and accountability. Maintaining continuity of personnel on a long-term project is difficult and would likely require additional investment in human resources. Reclamation appears to operate on the principle of collaborative or shared management centering on the regional directors. Although shared management can tend to prolong decision making, it can also function fairly well.

Centralized versus Decentralized Authority and Responsibility

Organizations can and do take on many forms, with varying degrees of success. Some will function successfully despite the form; others will falter under the best of theoretical forms. The internal culture and history of the organization play a significant role in determining the appropriate

---

[14]Based on information provided to the committee through August 2005. Subsequent to the committee's last meeting and development of this report, Reclamation undertook a reorganization that included a change in the organization of the Policy, Management, and Technical Services; Security, Safety, and Law Enforcement; and the Dam Safety Office. These changes were not completed in time for the committee to assess their impact or discuss them in this report.

structure and the ultimate outcome. Additionally, as is the case for Reclamation, pressures to reduce the federal workforce and increase the proportion of outsourced activities will continue to dictate changes in the structure and functioning of federal organizations.

The issue at the center of Reclamation's potential organizational changes involves centralization versus decentralization of authority, responsibility, and resources. As mentioned above, in the mid-1990s, Reclamation undertook a major reorganization to create a more decentralized structure (USBR, 1993). The effort was driven by, among other things, a change in the nature and quantity of the work, reductions in personnel and funding, and the goals of streamlining the organization, reducing administrative layers, and focusing the effort nearer to the site of the projects and Reclamation's customers. There is no question that benefits have been derived from this decentralization; however, there are also indications that problems have emerged. Over time, many organizations (private and government) having responsibility for facilities and infrastructure management have shifted from predominantly centralized, top-down management styles to various degrees and forms of decentralization. Some organizations have found their decentralization efforts to be either too extensive or carried too far down the chain of command, with the consequent loss of owner control. As a result, there has been some retrenchment from the belief that decentralization, in and of itself, is a panacea for producing efficiencies or satisfying customers and sponsors. Decentralization is plagued by a tendency to narrow the focus of the participants and to devalue legitimate organization-wide interests.

A major factor in achieving the desired balance between decentralized and centralized authority and responsibility is the quality and quantity of communication—particularly face-to-face communication. A lot can be achieved if managers at the area, regional, and headquarters levels know and trust each other. This trust is the product of consistent and open lines of communication. Without good communication, suspicions will grow and the organization will not function well. This means that for Reclamation to operate as a decentralized organization it needs to plan and budget for frequent meetings to exchange ideas on management and technical issues. It may be tempting to label such meetings as "unnecessary travel" and to cut the funds for them, but they may be among the most necessary activities in the travel budget. Absent a commitment of time and resources, the desired level of communication is not likely to take place.

Reclamation, like other customer-oriented agencies, needs to consider several factors that affect the optimum balance between centralized and decentralized operations:

- Retention of a close and continuous working relationship with local water users and other stakeholders in the project area.
- Customer and stakeholder preference for a strong, empowered area office.
- Stakeholder and contract partner concern for the cost of Reclamation services and decisions that affect their interests.
- Budgetary pressures that require ever-increasing efficiency in administrative and support functions.
- Younger employee expectations about empowerment and aversion to centralized control.
- Customer and stakeholder demand for agency consistency.
- Availability of expertise in critical technical fields and specialties at appropriate levels of the organization.
- Ability to effectively outsource nongovernmental activities.
- Personnel recruiting and development and the retention of core competencies.
- Effective and unequivocal delegation of authority and responsibility for key technical and administrative decisions.

The pattern best suited to administrative support may not be best for customer relations in the field. Close and continuing contact between local water users and Reclamation representatives in the field is essential to cooperative relations and in some instances to an adaptive-management approach to decision making. While a decentralized approach appears to address this need, unrestrained decentralization may lead to inconsistency. Decentralized responsibility accompanied by commensurate authority, defined and constrained by centralized policy, would therefore appear to be best suited to this scenario.

Administrative and technical support, unlike customer relations, would be amenable to a much stronger degree of centralization. In this time of instant electronic communication, there is little reason to expect problems with carefully managed centralized administrative support for many common functions. However, determination of the appropriate functions and the degree to which they are centralized requires judgment. Bureauwide centralization may well be justified in some cases, while regionwide concentration of activity might be more appropriate in others. On principle, administrative and technical support should be considered for centralization at the highest level that assures timely and effective response to field needs.

The committee believes that the following broad assignment of functions addresses the centralization versus decentralization question appropriately. The roles of the Commissioner's Office, deputy commissioners' offices, and PMTS are combined because all have a bureauwide focus.

Commissioner's Office, Deputy Commissioners' Offices, and PMTS

- Assume responsibility for communicating the bureau's mission and establishing strategies to accomplish it.
- Determine and promulgate policy.
- Rule on appeals of regional director decisions if necessary.
- Maintain contact and liaison with the secretary of the interior, other federal agency heads, and Congress.
- Speak for the bureau to the media on broad issues.
- Set Reclamation-wide priorities, including budget allocations.
- Select and supervise key personnel at the headquarters staff and regional director's level.
- Oversee major acquisition and high-risk projects.
- Determine core competencies for bureauwide activities.

Regional Offices

- Assume principal responsibility for facility engineering and resource management within the region.
- Assume principal responsibility for the construction processes and support to area and project offices on contract administration.
- Represent the bureau to state and local government officials, regional directors of other federal agencies, local media representatives, and user group officials, as appropriate.
- Rule on appeals of area and project manager decisions if necessary.
- Select and supervise key personnel at the regional, area, and project levels.
- Formulate and submit regional budget and recommend priorities.

Area Offices

- Serve as the principal point of contact with local water users, contract partners, local officials, and other stakeholders.
- Collect and submit field-derived engineering data.
- Recommend budget and priorities applicable to the area.
- Supervise O&M-related construction projects not assigned to a separate project office, including quality assurance, and ensure that contractors execute their quality control responsibility.
- Select and supervise area office personnel.
- Exercise delegated authority of the contracting officer's technical representative.

Project Offices

Project offices should exercise the same responsibilities and authority as area offices, but only for their own project. They should report to the regional director but coordinate with appropriate area managers. They should only have contact with the sponsors and users of their project. The extent to which project offices are self-sufficient administratively and technically will be determined by the regional director based on the stakeholders, scope, location, and duration of the project.

Technical Service Center

TSC, the largest element within the PMTS directorate, is somewhat analogous to a large engineering firm performing facility and infrastructure engineering design. A centralized Reclamation design organization that has a worldwide reputation for excellence in the water resources field has existed in Denver for many years, albeit in different forms. Although the FY 1994 reorganization shifted some work to area and regional offices and resulted in a smaller TSC, the unit has retained most of its technical competencies. At the same time, it has used benchmarking against private sector architecture and engineering organizations of similar type and size to streamline its business and management practices.

Despite TSC's long history and having been in place in its present form for nearly a decade, the committee heard comments from various stakeholders and to a lesser extent from Reclamation field units about inconsistent performance at TSC. The dissatisfaction centers on the following issues:

- A perception that the charges for services rendered exceed those that would be charged by the private sector or Reclamation field units.
- Excessive time required to complete projects.
- Overly stringent design standards in some cases.
- Insufficient responsiveness to customer views.
- Inconsistent competency and performance.
- Unnecessary personnel charging time to projects and attending project meetings.
- Retention of work that could be completed more efficiently by sponsors of transferred works.

The committee is in no position to verify or refute these perceptions without having access to an in-depth analysis of costs, schedules, and design performance. While the complaints may or may not be valid, the committee sees a continued need for a centralized design capability within

Reclamation. To be effective, it needs to have critical mass for efficiency and for sustaining the requisite technical competencies. Also, it is the only unit in Reclamation able to provide independent and consistent technical oversight of work done at the area and regional offices. However, unless there is clearer direction and support from senior management and closer coordination with the regions, TSC risks being considered irrelevant.

There are multiple centers of engineering and design expertise within TSC for various disciplines and specialties that undertake similar types of projects. The committee believes such capabilities should continue to be collocated to provide efficient collaboration rather than dispersed in communities of specialized practice throughout the bureau—that is, TSC should continue to be the source of the highest level of engineering and science expertise, and distributing design expertise to the regional offices would further degrade consistent implementation of policy and oversight of the process. Capabilities for more routine O&M, repair, and modernization projects should continue to reside at the regional or area office.

The committee carried out a high-level review of the TSC structure. It observed that many TSC units have similar functions and could be merged or even eliminated. Others that appear to only intermittently be of service to Reclamation should be reviewed. The TSC organization chart includes 39 functional units in five divisions, which appears to be excessive. Generally, an organization with too many organizational units incurs additional supervisory and administrative costs and keeps individuals from being assigned multiple tasks. The effect is a less productive organization.

Design, Estimating, and Construction Office

The Design, Estimating, and Construction Office (DEC) was recently established within the Operations Office for the purpose of instilling a consistent approach to the design, estimating, and construction functions, an approach that is missing in the present decentralized model. DEC is intended to fulfill some of the functions inherent to an owner's role in project management, and the committee commends the move in this direction. However, the committee is concerned that DEC appears to have limited authority and that its procedures do not appear to be thoroughly planned. The functioning of this unit should be evaluated as it progresses to ensure that it has the ability and the means to see that its findings and recommendations are given appropriate consideration. The committee believes that locating the office within the PMTS directorate is appropriate. An owner's role in project management and the role of DEC in improving project management in the bureau are discussed in Chapter 3.

Research

Reclamation conducts a research program to improve its ability to better manage water and power. The Reclamation Web site notes that "Reclamation conducts research to develop and deploy successful solutions for better water and power management—not to merely publish. Research is a vital paradigm for Reclamation, as Reclamation promotes rapid deployment of new innovations to benefit water and power operations" (USBR, 2005a). Research and Development is a unit under the PMTS director and is a parallel unit to TSC. Research activities include science and technology, desalination and water purification, and technology transfer. Research is conducted both in-house and by contract. Most of Reclamation's in-house research is undertaken by scientists and engineers in TSC and Reclamation's Water Quality Improvement Center (WQIC) in Yuma, Arizona. The research at the WQIC is focused on desalinization and water treatment.

Reclamation conducts research in the following areas:

- Water and power infrastructure reliability and safety,
- Water delivery reliability,
- Reservoir and river operations decision support,
- Water supply technologies, and
- Related environmental topics.

The committee supports the goal of expanding interagency research programs and believes that a good model is the Watershed and River System Management Program (WARSMP) sponsored by Reclamation's Science and Technology Research Program and the U.S. Geological Survey's (USGS's) Water Resource Division (USGS, 1999). This program developed a decision support system framework to assist water managers in making complex decisions. WARSMP included collaborative research with the Tennessee Valley Authority (TVA), the Department of Energy's Western Area Power Administration (WAPA), and the National Oceanic and Atmospheric Administration (NOAA). Reclamation's Water 2025 program is consistent with a cooperative approach to research and development.

Although there are several successful programs, the committee questions the justification for a research and development office separate from the research units within the TSC. While to a certain extent the work of the research office, such as research projects on desalinization, is basic research (as opposed to the research conducted within the TSC, which is more project related), this fact may not justify parallel organizations. Without an exhaustive review, the committee is in no position to make a judgment on this issue, but it does advocate that Reclamation consider conducting such a review to identify opportunities for increased efficiency. As for the larger issue of maintaining a laboratory facility, the committee questions whether such a facility is affordable or whether private, academic, and other governmental facilities could perform the work in a more cost-effective manner. This question becomes a question of whether all or part of the laboratory is necessary to fulfill the foreseeable mission of Reclamation. Further study appears warranted.

International Affairs

Reclamation's International Affairs Program within PMTS conducts a number of activities, including technology exchange, training, and technical assistance. The program's objectives are to "(1) further U.S. foreign policy, (2) enhance public health or promote sustainable development in developing countries, (3) support U.S. private sector participation in the international marketplace, and (4) obtain improved technology for the benefit of Reclamation water users and the United States" (USBR, 2005b).

Reclamation and other U.S. water resource agencies (USACE, TVA, USGS), as well as other institutions and companies in the United States, have long been esteemed worldwide for their accomplishments and expertise in this area. Reclamation's International Affairs Program has been the vehicle for sharing the bureau's expertise through training and technical assistance. The committee has, however, observed a significant reduction in Reclamation's international activities. This is due in part to competition for limited resources, but there also appears to be a policy of disengagement. The committee believes that Reclamation's participation in international organizations dedicated to water resources and hydropower should be continued and that technical exchange with water resource managers in other countries should be encouraged.

Other Elements in the PMTS Directorate

As with TSC, the other five subdirectorates in PMTS operate as service units rather than as line management. The combined staff of the five units is roughly two-thirds that of TSC, with the Management Services Office being the largest. As with TSC, the committee sees value in analyzing the organizational breakdown and the positions allotted to assess opportunities for consolidation and competitive outsourcing. The committee's interest in consolidation of units stems largely from the belief that corporate control is more easily maintained with a flatter organization.

Dam Safety Office

The Dam Safety Office (DSO), although intimately involved with maintaining infrastructure, is located in Security, Safety, and Law Enforcement (SSLE). The committee views the DSO as a line organization having programmatic authority as well as responsibility and accountability for dam safety. This relatively small unit is essentially a management unit receiving engineering and inspection services from TSC and site data and construction services from the respective field units. Although the location of the unit within SSLE as opposed to PMTS could be questioned, the committee found no indication that the dam safety program was not being discharged appropriately. Also, being located in SSLE might conform more closely with the Federal Guidelines for Dam Safety.

Operations Directorate

Most of the Reclamation workforce resides within the Operations directorate. This is appropriate as the directorate is responsible for the execution and operation of projects. It is a line-type organization—the five regional directors report to the commissioner through his deputy—and reflects the reorganization implemented in FY 1994.

Delegations of committee members met with personnel in the regional offices as well as with area managers and with user groups and other stakeholders. (Appendix B contains a list of these meetings and the issues discussed.) The committee observed that the regions have different organizational structures, capabilities, and workloads. In general, the regions appear to be functioning well notwithstanding the usual challenges faced in this type of endeavor. The morale of personnel and their loyalty to Reclamation's mission is commendable. Each of the five regions has responsibility for sustainment of a large portfolio of facilities. The committee saw examples of excellence. However, in general, the regions will need to more aggressively evaluate their asset inventory, manage their assets, and engage in constructive relationships with customers and stakeholders if they wish to accomplish the following:

- Build the capacity of customers to accept transferred works where appropriate.
- Establish metrics to evaluate the effectiveness of O&M of assets whether managed by Reclamation or the customers.
- Develop plans to handle transferred works that have not been properly maintained.

Stakeholders and users were concerned that there is too little decision-making authority at the project level. They would like to see more, if not most, authority at the local level (area and project offices). This desire for decentralization of authority is understandable, but there are some inherent risks. Reclamation needs to ensure that offices being assigned more responsibility have the requisite talent to discharge that responsibility. Depending on the workload and budgetary and personnel constraints, there is a limit to the feasibility of assigning requisite talent to every office.

Another factor in the equation is the need for consistency. A concern of the committee is the design capability extant in the various area offices. The number of engineers in area offices varies, with some offices having only one or two people. Relying on the area engineers to handle all the specialties that may be involved in a project carries some risk. The committee believes it may be more efficient to consolidate planning and design efforts not outsourced or undertaken by TSC in the regional offices. This will become more critical if further retrenchment in workload and workforce occurs or the type of workload changes materially.

REFERENCES

Achterberg, David. 1999. *Bureau of Reclamation's Dam Safety Program.* USCOLD Lecture Series, Dealing with Aging Dams. Denver, Colo.: U.S. Society on Dams.

Feldman, Martha S., and Anne M. Khademian. 2000. "Management for inclusion: Balancing control with participation." *International Public Management Journal* 3(2): 149-168.

Keys, John W. 2005. "Maintaining and upgrading the Bureau of Reclamation's facilities to improve power generation, enhance water supply, and keep our homeland secure." Testimony before the U.S. House of Representatives Committee on Resources, Subcommittee on Water and Power. July 19.

Moore, Adrian T., Geoffrey F. Segal, and John McCormally. 2000. *Infrastructure Outsourcing: Leveraging Concrete, Steel, and Asphalt with Public-Private Partnerships.* Reason Foundation. Available at http://www.acec.org/advocacy/pdf/fullstudy.pdf. Accessed November 16, 2005.

National Research Council (NRC). 2000. *Outsourcing Management Functions for the Acquisition of Federal Facilities.* Washington, D.C.: National Academy Press.

NRC. 2004. *Adaptive Management for Water Resources: Project Planning.* Washington, D.C.: The National Academies Press.

Office of Management and Budget (OMB). 2003. "Performance of commercial activities" (Circular No. A-76). Washington, D.C.: Executive Office of the President.

U.S. Bureau of Reclamation (USBR). 1993a. *Report of the Commissioner's Program and Organization Review Team.* Washington, D.C.: Department of the Interior.

USBR. 1993. *Blueprint for Reform.* Washington, D.C.: Department of the Interior.

USBR. 1998. "Review/examination program for high- and significant-hazard dams" (FAC 01-07). *Reclamation Manual.* Available at http://www.usbr.gov/recman/. Accessed August 16, 2005.

USBR. 2000. "A brief history of the Bureau of Reclamation." Available at http://www.usbr.gov/history/briefhis.pdf. Accessed August 16, 2005.

USBR. 2004. *Decision Process Team Report: Review of Decision Making in Reclamation.* Washington, D.C.: U.S. Bureau of Reclamation.
USBR. 2005a. "Bureau of Reclamation—about us." Available at http://www.usbr.gov/main/about/. Accessed August 16, 2005.
USBR. 2005b. "Bureau of Reclamation—international affairs." Available at http://www.usbr.gov/international/. Accessed August 16, 2005.
USBR. 2005c. *Reclamation Manual.* Available at http://www.usbr.gov/recman/. Accessed August 18, 2005.
U.S. Department of the Interior (DOI). 2005. *Orientation to the U.S. Department of the Interior.* Available online at http://www.doiu.nbc.gov/orientation/bor2.cfm. Accessed August 1, 2005.
U.S. Geological Survey (USGS). 1999. *The Watershed and River System Management Program.* Available at http://wwwbrr.cr.usgs.gov/warsmp/. Accessed August 16, 2005.

3

Good Practice Tools and Techniques

INTRODUCTION

The committee was asked to identify good practice tools and techniques for Bureau of Reclamation efforts in facility and infrastructure management. This chapter addresses practice tools and techniques in asset management, acquisition, and contracting, human resources, project management, and planning and budgeting that could be usefully applied to meet Reclamation's mission needs. The policies and procedures necessary for putting these good practices in place are also discussed. Tools and techniques for human resource management are reviewed in detail in Chapter 4.

Some of the committee's observations regarding practice tools and procedures stem from a roundtable discussion held with senior representatives from organizations with missions similar to Reclamation's, as well as current tools and techniques used at Reclamation that are viewed by the committee as representing good practices. Others reflect the committee's knowledge of and experience with the project management, acquisition, and contracting practices of other agencies.

ROUNDTABLE OF ORGANIZATIONS WITH SIMILAR MISSIONS

On June 22, 2005, the committee convened a meeting to discuss organizational and operational models used by other federal agencies and other governmental organizations with mission responsibilities similar to Reclamation's to identify good practice tools and techniques. Representatives of the U.S. Army Corps of Engineers (USACE), the Tennessee Valley Authority (TVA), and the California Department of Water Resources (DWR) participated in the discussion. The focus of the discussion was the facility and resource development and management practices used by these organizations. Following is a brief summary of the discussion and the committee's conclusions. Detailed notes are provided as Appendix C.

The committee observed that although the participating organizations had many similarities they also differed significantly in the size, scope, and focus of their missions. The three organizations also had differing cultures that support their unique methods of doing business. Although these differences inhibit the direct transplantation of policies, procedures, and organization, there are general lessons to be learned. Because of the relatively large size of the territory in which they operate, the operations of USACE seemed more analogous to those of Reclamation as a whole, while TVA and DWR can more readily be compared to Reclamation's regions.

U.S. Army Corps of Engineers

USACE's civil works mission is very similar to Reclamation's. The main difference is that Reclamation's operations are focused in the western states and USACE operates throughout the country. Reclamation focuses more on providing hydroelectric power and water for irrigation than USACE, which, while generating more hydropower than Reclamation, focuses more on flood control and navigation. Both organizations have had major construction programs to develop dams and waterways and are now responsible for the operation, maintenance, repair, and modernization of these facilities. All projects are undertaken with appropriated funds, but projects that require cost sharing are not implemented until sponsors secure their matching contributions.

USACE is composed of 41 districts, each having a fairly high degree of autonomy. The districts are organized into eight regions. Current mission requirements are driving USACE toward more uniform policies, procedures, and service to customers and are addressed by reducing autonomy and increasing central promulgation and local implementation of policies.

Tennessee Valley Authority

TVA's overall mission is to generate prosperity for the Tennessee Valley. There are three goals. These goals encompass requirements for maintaining navigation and flood control, established in the initial TVA legislation, as well as managing the Tennessee River system for aquatic habitat, water quality, water supply, and recreation. TVA's power system comprises 11 fossil fuel plants, 3 nuclear plants, 29 hydro plants, 1 pumped-storage plant, 6 combustion turbine plants, 7 diesel units, 16 solar energy sites, and 1 wind energy site. TVA is both a power producer and power marketer, and it operates as a federal corporation.

TVA, the nation's largest public power provider, serves 8.5 million residents and 650,000 businesses and industries. In addition to its ratepayers, TVA has many public and private stakeholders who are affected by how TVA manages the Tennessee River and TVA facilities and infrastructure.

California Department of Water Resources

DWR has about 2,500 employees, which is considerably fewer than either USACE or TVA. DWR has different constraints, but it also faces many of the same issues. DWR's mission is "to manage the water resources of California in collaboration with others to benefit the state's people and to protect, restore, and enhance the natural and human environment." Over 50 percent of DWR's personnel are assigned to the State Water Project (SWP), which covers much of the same geographic area as Reclamation's Central Valley Project (CVP) but is smaller and serves more urban customers. SWP includes 17 pumping plants, 8 hydroelectric plants, 30 storage facilities, and 693 miles of canals and pipelines.

Implications of USACE, TVA, and DWR Practices for Reclamation

Mission

Fifty years ago water projects were about economic growth and development. In the last 20 years environmental issues have grown in importance. The result is mitigation projects resulting from past decisions that did not recognize environmental issues and the current incorporation of environmental concerns in all engineering endeavors. Addressing the environmental aspects of the mission becomes a question of costs and benefits, and who pays. The public wants environmental protection but is often not willing to pay for it. The beneficiaries of water systems and hydropower experience increasing costs due to environmental conservation even as they receive a constant or diminished level of benefits. The issue is whether environmental conservation is a broad public benefit to be paid for by all or an integral part of the cost of hydropower, water, and flood control. The cost of increased security presents a similar problem.

Funding

A large subsidy is provided to agricultural irrigators through Reclamation's cost recovery limits. DWR passes on actual costs and provides no subsidy to its customers, but it provides assistance grants to local governments. These grants are used at the discretion of the local governments with few state requirements.

As a federal corporation, TVA is self-funded through rates paid by its electric power customers that cover all operating expenses. TVA's rates are also guided by its mission to generate prosperity. A cost recovery model is also used by the Reclamation units that operate large hydroelectric facilities off-budget. USACE, like most of Reclamation, relies on appropriated funds and cost sharing or reimbursement by the beneficiaries.

Finding funds to cover the costs of maintaining, repairing, and modernizing the nation's water-related infrastructure (hydropower, irrigation, municipal and industrial (M&I) water, flood control, and related facilities) is a problem faced by the federal government as well as state and local governments. Many projects were built with federal funds but rely on local initiative for maintenance. The cost of recapitalizing facilities and infrastructure that have exceeded their service life is often beyond local means, and there is no clear resolution of who should pay. The state of California is considering a water resource investment fund with funds collected from all water users throughout the state. Some of the fund would be controlled locally, where it is collected, and some spent statewide on broader needs.

Working with Sponsors and Stakeholders

Managing water systems requires a highly collaborative process. It requires coordination among government agencies at the federal, state, and local levels and coordination among water users and other stakeholders, and between government agencies and users and stakeholders. Water agencies need engineers who can collaborate with others. They are expected to work across disciplines and with the public. DWR does not specifically evaluate this capability in its personnel, but it is nonetheless a critical part of its success.

USACE finds that it increasingly plays the role of facilitator. As the group responsible for managing the water systems, it needs to bring the users and stakeholders together to identify issues of common concern, areas of agreement, and issues that need to be resolved. Planners are trained in facilitation skills and expected to take a leadership role in applying dispute resolution systems. The engineering solution is often secondary to the resolution of divergent public interests.

Performance reviews of managers and specialists at TVA are divided into two parts. Seventy percent is tied to measurable performance goals and 30 percent is tied to behaviors. The ability to collaborate is a desired behavior. In the past, TVA tried to take the responsibility for finding a fair solution by understanding and representing all interests. This placed TVA in a position where it was at odds with most stakeholders. The situation has been improved by stepping back and letting the interested parties resolve their differences and then acting on the consensus decisions. TVA did not need to be a facilitator. Direct communication among members of the community made the difference. TVA also works collaboratively with the local community when stakeholders are ready to collaborate.

The key to effective relationships with sponsors and stakeholders is open and honest communication. The more transparent the agency's processes, the easier it is to get buy-in from sponsors and stakeholders. When sponsors and stakeholders do not agree, it is better for an agency to be neutral and allow them to arrive at an appropriate compromise.

Project Management

Completing construction projects within the original cost and schedule is a challenge for most organizations because of uncertainties in cost and schedule estimates. USACE addresses this challenge with a policy of adjusting designs to fit the budget unless the adjustments significantly alter the original scope. Cost estimating is particularly difficult in major rehabilitation projects when the nature and extent of existing conditions are not known until a portion of the project has been executed. Rehabilitation projects have a greater need for forensic engineers and institutional knowledge of how existing facilities are configured and operate.

Workforce Development

All organizations, especially those that rely on the technical competence of their workforce, are concerned with recruiting, developing, and retaining skilled personnel to maintain their core competencies. There are many tools and techniques used to achieve these objectives, all of which can be successful if used appropriately. USACE takes the long-term approach by addressing students in middle and high schools. TVA has found that retention is improved if they recruit from universities in their region. All three organizations have career development and training programs that include technical as well as managerial objectives. For Reclamation and the other organizations, maintaining the commitment and funding to implement their chosen programs can be a problem.

USACE has determined that an average new graduate engineer will have eight jobs in his or her working career. In USACE this means about one-third of the workforce will have a tenure of 8 to 10 years. USACE believes that the federal benefit package makes it competitive with the private sector in attracting and keeping qualified personnel. The challenge, for any organization that is project driven, is dealing with the variations in demand. USACE also invests in its human resources by giving its engineers 40 to 80 hours of training per year. This includes technical as well as management training. It is important to select people who have the traits needed by a manager for management positions and to find other ways to reward people who are better suited to technical positions. In USACE this applies to all disciplines employed in the organization. In recent years USACE has also recognized project management as a discipline. DWR has a target of about 50 hours per year of training and also supports efforts by its employees to earn advanced degrees.

Another challenge is to retain the institutional knowledge possessed by people who are retiring. USACE does this by conducting extensive exit interviews with all retirees and recording the resulting information in a database. Downsizing over the last few years has reduced the opportunities for mentoring whereby senior personnel can pass their wisdom on to the new people in the field. Institutional mechanisms are needed to formalize this transfer of knowledge. USACE has a rotation program for new hires; DWR does not.

USACE is looking at bringing more senior engineers from outside the organization into leadership positions. There is some internal bias against this, but it can be overcome. USACE is applying some effort in middle and high schools to promote careers in engineering and in the corps. This same approach needs to be applied to the O&M crafts as well.

TVA is targeting its recruitment at the best and the brightest in the South. This geographic focus is reducing the pool of potential recruits but increasing the hiring success and retention rate.

Centralized versus Decentralized Engineering Services

The geographic area of responsibility for TVA and DWR is roughly equivalent to that of Reclamation regional offices. TVA and DWR both rely on centralized engineering design services and dispersed facility inspection and maintenance. Automation has also allowed centralization of many operation functions. Personnel in the field act as the owners, while central office personnel provide consulting services. As the owners, field personnel maintain control of the process.

Large agencies like USACE and Reclamation have a need for consistency throughout the organization. This can be achieved by centralized operations or the development and implementation of strong guidelines and standards that are developed centrally but implemented locally. This does not preclude local participation in the development of policies and procedures.

In-House versus Contract Services

Since the 1990s, there has been a sustained effort in government to reduce the quantity of services performed by government employees and increase contractor-provided services. To function as a smart buyer, an organization that requires technical services often retains a minimum level of technical expertise in-house in order to select and manage outside contractors effectively. There is also general agreement on the necessity of undertaking technical activities in order to maintain the expertise needed to manage contractors. The problem is then determining the optimum mix of in-house and contract services.

The application of arbitrary targets for the quantity of contract services can be problematic. As noted by TVA, outsourcing decisions are based on availability and economic factors. It should be recognized by those who would increase reliance on contracting that it is often more difficult to regain core competencies after they are lost than to maintain them.

Impact of Environmental and Social Issues

Environmental and social issues are an integral part of all water projects. The development process should integrate these issues from the beginning, even if they increase the final cost. Addressing them as an add-on after the fact is even more expensive. Most current problems are the result of past failure to recognize their importance. The issue that remains to be resolved is whether the costs of mitigating the environmental and social consequences of water projects are to be paid by the direct beneficiaries of the water and power projects or by the general population.

POLICIES AND PROCEDURES

As noted in Chapter 2, the committee believes that existing policy, as promulgated in the *Reclamation Manual*, to guide Reclamation's decisions and actions for the benefit of stakeholders, employees, and the public at large is inadequate. USACE reported that it is facing some of the same problems but is reacting as follows:[1]

---

[1] Donald Basham, U.S. Army Corps of Engineers, Remarks at roundtable discussion on June 22, 2005.

- Driving toward more consistency nationwide.
- Establishing more standard procedures and processes.
- Focusing increasing responsibility on the regional organization as opposed to the Washington, D.C., headquarters or the geographically dispersed districts.
- Centralizing selection of key senior employees to promote consistency.
- Utilizing standard designs where possible.
- Increasing involvement with project sponsors and stakeholders at all stages of project planning and design.
- Using centralized guidance with local implementation.

The USACE chief of engineers, LTG Carl Strock, in a message to all corps employees dated April 13, 2005, said the following:[2]

There are four non-negotiable aspects to the USACE 2012 strategic plan,

- We will act as one headquarters to streamline our processes. This is not a structural combination, but rather a unity of effort. By combining the efforts of the Washington D.C. headquarters with the division headquarters, we reduce a layer of review and therefore, improve the timeliness of actions.
- We will have regional integration teams in Washington D.C. to focus on supporting Regional Business Centers.
- We will have Regional Business Centers that share resources throughout the region and multiply our capabilities.
- We will maintain active Communities of Practice to help us maintain technical competence and share knowledge. My intent is for us to achieve a level of national consistency so employees can move to any district and know the processes and procedures.

TVA also reported that it is moving toward standardizing more processes and procedures.

A key factor that appears to be driving agencies toward centralized policy promulgation and service support is budget pressure. All agencies report pressure to do more with less. The commissioner recognized this principle in the FY 1994 reorganization, but the absence of policy guidance and the decentralization of much authority have made the promised efficiencies harder to achieve.

---

[2] Message from LTG Carl Strock, chief of engineers, U.S. Army, to all corps employees, Subject: Keeping You Informed, dated April 13, 2005.

ACQUISITION AND CONTRACTING PRACTICES

The following sections address acquisition and contracting good practices, including ways to determine whether Reclamation activities should be undertaken by government or contractor personnel and ways to ensure that Reclamation staff can be made aware of innovative and effective contracting approaches.

Competitive Sourcing Policies and Practices and the Level of Outsourcing

The bureau relies on its own regional and area employees, TSC staff, and contractor support to meet mission needs. While the days of huge new dam construction projects appear to be over, there is still a strong need for solid technical and engineering expertise to deal with the many infrastructure issues associated with Reclamation's aging facilities. However, responsibility for O&M for a number of these sites has been shifted from the bureau to the local water districts. As noted in Chapter 2, this is the case for 428 of Reclamation's 673 facilities. A comment heard by the committee is that water customers believe the bureau charges considerably more for projects and for Reclamation-performed work than do the districts and private sector consulting and engineering firms.

There appears to be no set bureau policy about when to obtain contractor support and when to look to internal staff to do the work. Decisions of this sort are made at the region and area levels or at TSC, as opposed to by headquarters. Frequently these decisions appear to be based on the availability of in-house staff to conduct the work.

Most of the O&M-type work conducted by the bureau would by no means be considered inherently governmental. Therefore, virtually all of this work could be contracted out, using private sector capabilities and allowing the bureau to reduce staff and costs.

The Office of Management and Budget (OMB) Office of Procurement Policy (OPP) Policy Letter 92-1 of September 23, 1992, originally established the government-wide policy for addressing inherently governmental functions. The thrust of this policy can now be found in the Definitions section and subpart 7.5 of the Federal Acquisition Regulation. The basic definition of an inherently governmental function from Policy Letter 92-1 is as follows:

> As a matter of policy, an 'inherently governmental function' is a function that is so intimately related to the public interest as to mandate performance by Government employees. These functions include those activities that require either the exercise of discretion in applying Government authority or the making of value judgments in making decisions for the Government.

The policy explicitly noted building maintenance as a function that could be performed by contractors. Although the committee recognizes the difference between O&M of buildings and the O&M of Reclamation facilities, one could easily construe the definition in Policy Letter 92-1 to cover other types of maintenance and support work as well.

The committee believes that the National Research Council report *Outsourcing Management Functions for the Acquisition of Federal Facilities* offers a good model for Reclamation to follow as it makes its determination of inherently governmental functions related to its infrastructure activities. The following section from the Executive Summary of that report describes the approach:

> Although design and construction activities are commercial and may be outsourced, management functions cannot be clearly categorized. In the facility acquisition process, an owner's role is to establish objectives and to make decisions on important issues. Management functions, in contrast, include the ministerial tasks necessary to accomplish the task. Based on a review of federal regulations, the committee concluded that inherently governmental functions related to facility acquisitions include making a decision (or casting a vote) pertaining to policy, prime contracts, or the commitment of government funds. None of these can be construed as ministerial functions. The distinction between activities that are inherently governmental and those that are commercial, therefore, is essentially the same as the distinction between ownership and management functions.
>
> Using Section 7.5 of the Federal Acquisition Regulations as a basis, the committee developed a two-step threshold test to help federal agencies determine which management functions related to facility acquisitions should be performed by in-house staff and which may be considered for outsourcing to external organizations. The first step is to determine whether the function involves decision making on important issues (ownership) or ministerial or information-related services (management). In the committee's opinion, ownership functions should be performed by in-house staff and should not be outsourced.
>
> For activities deemed to be management functions, the second step of the analysis is to consider whether outsourcing the management function might unduly compromise one or more of the agency's ownership functions. If outsourcing of a management function would unduly compromise the agency's ownership role, then it should be considered a "quasi"-inherently governmental function and should not be outsourced. (NRC, 2000, pp. 3-4)

Policy Letter 92-1 cautions that other factors also may play a role in that decision. It states as follows: "Determining whether a function is an inherently governmental function often is difficult and depends upon an analysis of the factors of the case." Along these lines, it points out the need for agencies to maintain a core capability in key disciplines whether commercial or not to ensure that the government remains a knowledgeable and informed buyer of contracted services. The policy states as follows:

> Agencies must, however, have a sufficient number of trained and experienced staff to manage Government programs properly. The greater the degree of reliance on contractors the greater the need for oversight by agencies. What number of Government officials is needed to oversee a particular contract is a management decision to be made after analysis of a number of factors. These include, among others, the scope of the activity in question; the technical complexity of the project or its components; the technical capability, numbers, and workload of Federal oversight officials; the inspection techniques available; and the importance of the activity. (OPP, 1992, p. 7)

In other words, an agency may well need to maintain proficiency in what otherwise would be commercial activities to ensure it remains an informed buyer of such services. Area, regional, and Denver PMTS staff have pointed out that maintaining an internal capability allows them to address precisely this issue. In this connection, Reclamation should train its contracting officer's technical representatives (COTRs) to ensure that they possess the skill sets necessary to oversee that contracted work.

Effectiveness of Contracting Techniques and Methods

Although Reclamation uses a variety of procurement methods for construction, the bulk of construction work is procured by firm fixed-price contracts through sealed bidding or negotiation. In the case of source selection, awards are made on a best-value basis. Design-build contracting is being considered but has not yet been used to any great extent. Most invitation for bids and request for proposal procurements are set aside for small businesses, businesses owned by minorities, females, and other disadvantaged persons, and historically underutilized business zones (HubZones) unless it is determined that the capabilities or the competition is inadequate. In that case, full and open competition is used.

Over the last decade federal agencies have adopted a variety of acquisition techniques to streamline and improve contracting performance. These practices have relied on new contracting vehicles such as the General Services Administration (GSA) schedules, tasks under multiyear indefinite delivery/indefinite quantity (IDIQ) contracts, or simplified acquisition of basic engineering requirements (SABER) to streamline procurements while still addressing agency mission needs and inspiring adequate competition. Reclamation has used these techniques to meet a variety of contracting needs. For example, the Lower Colorado region is obtaining good results with IDIQ contracts for its maintenance and repair work. Under an IDIQ, a contractor bids on tasks associated with a particular contract since it has already gone through a full and open competitive process and has been issued an award. This technique greatly speeds up and simplifies the contracting process.

In addition, the region is now developing contracting approaches to be used in its new Multi-Species Conservation Program, which is a 50-year effort totaling over $600 million. There will be a 50 percent nonfederal cost share for this effort and 40 nonfederal permittees. For this project a 35-member steering committee involving three states will be established, and five chairs will be rotated among 40 customer representatives. Since a number of Reclamation projects require both cost sharing and nonfederal-stakeholder participation, the committee believes that the bureau should develop a series of contracting templates so that all regions can take advantage of the approaches followed and lessons learned by various regional and area offices.

A contracting technique that is now enjoying widespread use across the government is performance-based services acquisition (PBSA), which requires an agency to identify desired business outcomes but allows the contractor to use its own methods to obtain these results. Staff at some of the regions have described using this approach for acquiring relatively low-level services—for example, janitorial support for Hoover Dam. However, Reclamation staff should explore further the use of this technique to focus contractor-provided maintenance support more on the bureau's desired business outcomes. Moreover, PBSA approaches can be applied to many different types of service contracts, including those for high-level professional and technical services. Clear performance measures are a means of monitoring whether the contractor is performing successfully.

Some of Reclamation's acquisition staff have used innovative contracting methods. For example, staff in the Pacific Northwest are using a reverse auction approach to achieve significant financial savings. Under a reverse auction, vendors bid to lower the prices for the commodities to be purchased. This is a technique that has been widely used by the commercial sector but less so by government. The committee encourages such innovative contracting approaches as a way to get the lowest prices and the best value for the bureau.

In addition to Reclamation contracting under federal acquisitions regulations (FAR), some contracts are executed by water districts or by American Indian tribes under P.L. 93-638 authority. P.L. 93-638, the Indian Self-Determination and Education Assistance Act, was signed by President Ford in 1975. While initially directed at allowing tribes as sovereign nations to take over control of their own health-care programs from the Indian Health Service, the law has come to be used for a variety of purposes—among others, allowing a tribe to control delivery of services to its community, including contracting services. Various tribal contractors have made use of this authority in providing community construction and support services. In these cases, the selected entity executes the design and construction work under Reclamation oversight. The use of P.L. 93-638 authority is relatively new to design and construction, and results have been mixed. For example, the Animas–La Plata project in southwestern Colorado and northwestern New Mexico is being executed by Reclamation (design and construction management) and the Ute Mountain Ute tribe (construction contracting). This project now appears to be under control and headed for successful completion, but it experienced significant cost and schedule problems in its early stages.

The committee sees the benefit of this approach as the enhancement of tribal opportunities and American Indian self-determination. However, given concerns about the limited success of some projects employing this approach, the committee believes that significant up-front planning and sound project management and risk management analyses need to be performed to ensure that effective capacity and expertise are available. This is another area where best practices and lessons learned might be shared among bureau regions through some type of central contracting office Web site or repository.

PROJECT CONCEPTION, DEVELOPMENT, AND EXECUTION PRACTICES

Management of large construction projects is what Reclamation was all about at its inception and for much of the twentieth century. Accordingly, the Reclamation Instructions included a comprehensive set of policies and directives for planning and executing projects. As discussed above, action was taken in 1993 to sunset all such directives. Work has been under way since that time to redevelop a comprehensive set of procedures to provide consistency in project management throughout Reclamation.

Project Management Policies, Directives, and Guidelines

A Reclamation design and construction coordination team (RDCCT) was established in December 1996 to identify good practices for design and construction within Reclamation. The team comprises two members from each region and the Technical Service Center (TSC), one specializing in design and the other specializing in construction, plus two additional members from TSC. Among other things, they have developed the following policies, directives, and guidelines. The policies and directives are published as part of the continually updated, Web-based *Reclamation Manual* (USBR, 2005a):

Policy

- *Performing Design and Construction Activities*, February 11, 2000 (FAC PO3) and
- *Cost Estimating Policy Document* (under development) (FAC PO x).

Directives

- *Maintenance of Design and Construction Capabilities*, September 29, 2000 (FAC 03-01),
- *Construction Activities*, September 29, 2003 (FAC 03-02),
- *Design Activities*, July 9, 2004 (FAC 03-03),
- *Professional Registration for Engineers and Architects*, May 17, 2002 (FIRM 05-01), developed in conjunction with Human Resources (HR) and issued by HR,
- *Cost Estimating* (under development) (FAC 0X-01), and
- *Project Cost Estimate* (under development) (FAC 0X-02).

Guidelines

- *Final Design Process* (USBR, 2005a),
- Design data guidelines (about 80 percent complete as of May 2005),
- Drawing management portion of *Information Management Handbook*, Volume III, *Drawing Management and Drafting Standards* (USBR, 2000), and
- Other design standards in various stages of development.

In addition, some Reclamation Instructions that were cancelled by the sunset process are still being used as guidelines until replacements are in place. While technically not binding, they are used as a matter of good practice.

Project Management Practices

The referenced policies, directives, and guidelines establish the following as good practices for design and construction:

Responsibility

The regional director is given responsibility for accomplishing all project activities from initial appraisal planning through construction project closeout within his or her region (FAC PO3). The regional director, in turn, may delegate area managers or project managers to manage individual projects. The assignment of responsibility shapes the decision-making process and has a major impact on the ability to manage projects well. As indicated above, this process is under review to clarify lines of responsibility and accountability.

Programming

Annual work plans are developed for the majority of Reclamation projects as part of the annual Reclamation budget process. Project programming information includes project description, target schedule, and funding requirements by year, combined with funding justification (FAC 03-01). The Reclamation Budget Review Committee (BRC) reviews an annual zero-based budget to establish overall priorities. However, Hoover, Parker, Davis, Grand Coulee, and Glen Canyon dams are off-budget operations because all funding requirements are paid by power customers, and their projects are not prioritized by the BRC. The committee found that the procedures to develop and prioritize these projects are rigorous and well accepted by power customers and believes they should be used in other off-budget operations.

Project Planning, Authorization, and Cost Estimating

Procedures for preconstruction activities are contained in FAC 03-02 and FAC 03-03 and in TSC's project management guidelines. The TSC guidelines apply to TSC employees, but the principles can be used by any regional office. Requirements for establishing a project management team (PMT) and developing a project management plan (PMP) and descriptions and examples of components of a PMP are given in the manual.

Reclamation projects have three status categories: (1) planning (including appraisal, feasibility, and preliminary design studies), (2) construction (including final design), and (3) operations and maintenance. Within these categories there are two project stages for planning and four stages for construction. Cost estimates developed for each stage are prepared in increasing detail. Appraisal cost estimates are used to help Congress determine whether more detailed investigations of a potential project are justified. Appraisal estimates are not intended for requesting project authorization or construction funds from Congress. Feasibility cost estimates are based on information and data obtained during investigations for predesign and preliminary activity. The estimated costs of feasibility studies are part of the annual budget and must be authorized by Congress before the investigations begin. The construction cost estimate (CCE) and summarized project cost estimate (PCE) are normally prepared as part of the feasibility study. They complete the planning stage and are used to form the basis of the initial request for construction funds.

Reclamation's cost estimating procedures have been drafted but not yet published. The manual should include detailed procedures for establishing and controlling contingencies, and the certainty of the estimates needs to be linked to risk management procedures (FAC 0X-01).

The committee examined the Animas–La Plata (ALP) project to determine if there were any underlying flaws in the process that caused the problems encountered by the project. In 2003 the construction cost estimate increased from the 1999 level of $337.9 million to $500 million. A report to the secretary of the interior (USBR, 2003) concluded that while no single reason for the increase was found, there were several contributing factors, including reliance on inapplicable or incomplete data, inexperience with the cost impact of P.L. 93-638 contracting, and a decade of turmoil in defining the scope and deciding whether the project would be built. A Reclamation review of the original construction cost estimate found that it was not reliable, but the focus at the time was on completing environmental compliance and supporting efforts to reach internal agreement on a plan for the project, and the finding of unreliability was not followed up. Accordingly, the incomplete 1999 estimate was used by Congress to authorize the project in December 2000. The committee notes that a rigorous project management process, including extensive preproject planning and detailed cost estimating procedures, is usually the most effective means of developing reasonable cost estimates. Such a process did not appear to have been part of ALP. However, it appears that given the circumstances surrounding the Animas–La Plata project, the committee cannot be sure that an effective project management system could have prevented the problems encountered by ALP.

Design

Design work undertaken during the appraisal and feasibility stages is classified as planning. Final design begins at the start of construction status, following initial appropriation of project funds. The guideline, *Final Design Process* (USBR, 2005a), lays out a comprehensive set of criteria for design activities from the period before design data collection through construction to full operation of the facility. Its introduction notes that activities described therein may not be necessary for projects of limited scope. The guideline appears to cover the design process thoroughly and permits tailoring to meet the specific needs of the project.

Quality Assurance and Quality Control

Quality assurance and quality control (QA/QC) during the planning and design process are addressed in FAC 03-02 and FAC 03-03. They take the form of checking procedures, technical reviews, peer reviews, constructability reviews, and value engineering studies. The TSC Operating Guidelines (USBR, 2005c) outline these functions for TSC design work.

Quality control during the construction process is frequently part of a construction contract, but Reclamation maintains a cadre of construction inspectors in each region to assure quality construction. Some regions have used contractors for portions of construction inspection, contract administration, and materials testing work. The Mid-Pacific Construction Office, for example, has done so since 1994. FAC 03-02 outlines requirements for determining the extent of contractor quality control and bureau-independent quality assurance inspection and materials testing.

Project Closeout and Follow-up

Requirements for project closeout are listed in FAC 03-02 as post-construction activities. Included are contract closeout, preparation of as-built drawings, preparation of a technical report on construction, and design summary, designers' operating criteria, and O&M manuals. Transfer of project works from construction to O&M status is a formal process governed by FAC 01-05.

The technical report on construction and the design summary are the vehicles for documenting and passing on lessons learned throughout the acquisition process. Other lessons learned are promulgated by distribution of reports such as the Animas–La Plata cost estimate report mentioned above.

Reclamation's Role as Owner

The basic role of Reclamation as the owner of construction projects is to ensure that the bureau undertakes the right projects and executes them effectively and efficiently. However, creating and maintaining an organizational process that does this consistently is complex. In 2001, the Department of Energy (DOE) sponsored and NRC conducted a government/industry forum on the owner's role in project management and preproject planning (NRC, 2002). The forum presented case studies of how large organizations develop a project management culture and the steps they take to ensure that they undertake the right projects and execute them effectively. The forum examined the processes and procedures for developing buildings and industrial facilities, which are also applicable for developing and sustaining facilities for delivering water and power. The characteristics of owners of successful projects were described by the forum's organizing committee:

- Successful project management requires the institution of a project management discipline that encompasses all projects. It is not sufficient to do some projects well; what is needed is consistency. All the firms represented in the forum have well-defined, disciplined project processes, with buy-in and active participation by senior management.
- There is an absolute requirement for emphasis on project justification and identification of business or (in the case of DOE) mission need early in every project, even before a project is formalized. Senior corporate (agency) management must be closely involved in this process, as it is their responsibility to identify and interpret business or mission needs.
- Decision points with options for project approval, go-ahead, change, rework, or termination must be clearly identified. These decisions must be made by appropriate senior managers. The view that the need for senior management decisions slows down good projects is explicitly rejected. A good decision process actually expedites projects, in that it assures that they have the necessary resources, support, and direction to go to successful completion and operation—not merely to the next phase.
- Accountability and responsibility for project performance must be made clear and well defined across the enterprise. For the enterprise to succeed, all elements must succeed.
- A corporate organizational structure for project management must be established and maintained.
- There must be continual, formal project reviews by responsible management. Expectations, products, and metrics must be clearly defined for the entire process.
- There is no substitute for thorough front-end planning. This is true even better, especially for first-of-a-kind and one-of-a-kind projects. A successful project-management improvement process requires a cultural change, and cultural change is driven from the top. (NRC, 2002, p.viii)

All of the case studies emphasized the role of the owner in ensuring effective front-end planning activities. These activities include organizing the project team, evaluating and selecting options, defining the project in terms of quantity and quality, and establishing baseline budgets and schedules. The resulting product is called the project scope of work or project definition (NRC, 2003). The Federal Facilities Council (FFC) study found that "although preproject planning appears to be done thoroughly on some federal projects, the overall planning effort is inconsistent. Most of the agencies interviewed limit their preproject planning efforts, especially relatively costly activities, to major projects" (NRC, 2003, p. 2).

The Construction Industry Institute has collected data that link the quality of front-end planning to the success of projects. It has used these data to develop a process—the Project Definition Rating Index—for evaluating project planning to determine if a project is ready to proceed to final design and construction (CII, 1999). The tool was developed for buildings and industrial projects and has been adapted by DOE for environmental remediation projects. The committee believes that it could also be adapted for use on water and power projects.

Reclamation has recognized the need for high-level oversight of decisions and construction project management. As discussed in Chapter 2, the central Design, Estimating, and Construction Office (DEC) has been created within the Operations Office for this purpose. It will review projects costing more than $10 million, projects deemed to pose a substantial risk for the bureau, and other projects designated by the commissioner. The committee believes that DEC's oversight should also include front-end planning activities to ensure the accuracy and completeness of project scopes, risk management plans, and execution plans before projects proceed to design, because some of the problems of project schedules and cost estimates may be caused by deficiencies in the planning process. The committee also believes that Reclamation should establish criteria for the direct participation of the commissioner or his or her designated representative in project reviews. The level of review should be consistent with the cost and inherent risk of the project. The review procedures, processes, documentation, and expectations at each phase of the project need to be developed and applied to all projects, including those approved at the regional level.

CUSTOMER AND STAKEHOLDER RELATIONS

Since the establishment of Reclamation, the influence of customer and stakeholder input has evolved to the point that it has a significant impact on Reclamation's design and construction projects. Water districts, power customers, and Indian tribes have acquired expertise and experience and, in many cases, the ability and desire to do design and construction with their own contractors and consultants. For example, the Northern Colorado Water Conservancy District (NCWCD) held discussions with the Great Plains region about design and construction management services for a new Carter Lake Outlet, which is a part of the Colorado–Big Thompson (CBT) project. In this case the district wanted to undertake the design and construction with Reclamation oversight, but Reclamation determined that it was a high-risk project that should be under Reclamation control. The district is providing the funds for the project, but the facility is federally owned and Reclamation is liable for unforeseen consequences. Reclamation's approach is estimated to cost more, but the district's is more uncertain. The committee was impressed by the level and detail of communications on this controversy. Reclamation retained control of the project, but there appeared to an open exchange of ideas.

Another case involving the CBT and NCWCD concerned a Safety of Dams (SOD) project on the Horsetooth Dam, where conditions discovered during the course of the project allowed Reclamation to complete the project at considerable savings. While the total cost of the project was much less than expected, the costs of overhead and administration, as a percentage of the total costs, were higher than normal. Reclamation's cost reporting systems provided insufficient detail to explain the components of these expenses and why they were a significant part of the project costs. In this case Reclamation's project management procedures and communication with the district were not adequate.

An example of excellent stakeholder communications was observed by the committee at the Lower Colorado Dams Office. In the Lower Colorado region, power customers fund operation, maintenance, repair, and rehabilitation projects through their rates and have an oversight committee to review proposed O&M and rehabilitation plans. Reclamation develops 10-year O&M plans, which are reviewed by the oversight committee and become the basis for determining budgets and power rates. The Parker Dam generators' overhaul and upgrade is an example of how good stakeholder communications can work for the benefit of all parties. The power customers' oversight committee was concerned that an asset evaluation study by Reclamation would result in an overly conservative and therefore expensive program. The power customers requested that the study be performed by an independent contractor. The firm Montgomery Watson Harza performed a study that recommended turbine runner repair and generator rewinding projects, which are actually more extensive than originally proposed by Reclamation (MWH, 2002). The projects were approved by the oversight committee and were under way and appeared headed for successful completion at the time of the committee's site visit.

Research has shown that relational trust, which comes from a fair process and customers being treated with dignity and respect, is more important for the acceptance of policy decisions than instrumental, or calculus-based, trust[3] (Tyler and DeGoey, 1996). This research also supports the idea that "trust is a social commodity" that "gives authorities a 'cushion of support' during difficult times" and that it cannot necessarily be built

---

[3]Relational trust, also known as knowledge-based trust, is derived from interpersonal relationships. Instrumental trust, or calculus-based trust, is derived from a fear of the consequences of broken trust.

in the short term but needs to be nurtured and maintained (Tyler and DeGoey, 1996, p. 345). In other words, trust can be significantly enhanced by paying attention to how customers, stakeholders, and others are included in the process.

Reclamation works with a very broad range of customers and stakeholders, some of which have opposing objectives. While the committee has heard complaints about particular project issues and the decision-making process, in the end most Reclamation customers have a favorable view of Reclamation as a business partner. In situations where customers and stakeholders hold Reclamation in high regard, their positive feelings are based on trust developed with key Reclamation personnel. Extra care must be taken in selecting their successors to ensure that the quality of communications and level of trust are maintained.

APPLICATION OF METRICS, AUDITS, AND REVIEWS

In the case of the larger hydroelectric generating facilities, Reclamation uses an independent benchmarking process to determine how its facilities compare to others in terms of costs, reliability, efficiency, and overall maintenance. Such reviews are conducted on an annual basis, and the reports provide useful information to facility managers. Similar efforts should be made to establish metrics and measure the performance of Reclamation's water management assets. Reclamation regional offices reported the use of some review tools, including annual, periodic, and comprehensive facility reviews, value engineering reviews, and peer review of endangered species recovery programs. However, there seem to be wide differences in the application of such tools across the bureau.

The committee was informed that there are several forums within Reclamation to identify best practices for asset management, but the committee did not observe an effective dissemination of these practices. For example, in one region the issue of encroachment on Reclamation facilities by urban development and recreational uses was discussed, but no solution was suggested nor did any impact assessment information appear to be available.

PLANNING AND BUDGETING

The last two sections address the use of out-year budget planning documents by some parts of Reclamation to ensure stakeholder support for asset O&M or refurbishment needs and point out a problem with O&M funding that will likely increase unless steps are taken to deal with it.

5- and 10-Year Plans

The committee has observed effective systems for planning and executing O&M for facilities of various types and conditions. The core of the process consists of the 5- and 10-year plans developed in various regions to identify out-year funding requirements and to ensure that stakeholders are informed well in advance of future funding requirements, especially for refurbishment. However, the committee recognizes that the O&M burden for an aging infrastructure will increase and that the resources available to Reclamation, its customers, and contractors may not be able to shoulder the increased burden. This will challenge Reclamation to be more innovative and more efficient in order to get the job done. Given the success of the planning process in the Lower Colorado region, the committee believes that all regions should develop and use such plans as a stakeholder communications tool and as a roadmap for meeting future requirements. The committee believes that effective planning is the key to O&M of Reclamation facilities. In addition, Reclamation should identify the best practices for inspections and developing O&M plans and use them throughout the organization.

Funding for O&M Needs

A number of stakeholders pointed out to the committee the difficulties resulting from the requirement to reimburse expenditures for O&M activities within the fiscal year in which they were expended. This is a particular difficulty for some water districts that do not have enough control over cash flow and other factors to do this when O&M costs increase. Better long-term planning should allow these districts to anticipate such needs. However, if there are large spikes in required funding, it will still be difficult to meet this requirement in the limited time frame available. This problem will only become more severe as demand for O&M funding continues to grow.

REFERENCES

Construction Industry Institute (CII). 1999. *Project Definition Rating Index—Buildings*. Austin, Tex.: CII.

Montgomery Watson Harza (MWH). 2002. *U.S. Department of the Interior Bureau of Reclamation Parker Hydroelectric Powerplant Asset Evaluation Report*. Broomfield, Colo.: MWH.

National Research Council (NRC). 2000. *Outsourcing Management Functions for the Acquisition of Federal Facilities*. Washington, D.C.: National Academy Press.

NRC. 2002. *The Owner's Role in Project Management and Preproject Planning*. Washington, D.C.: National Academy Press.

NRC. 2003. *Starting Smart: Key Practices for Developing Scopes of Work for Facility Projects.* Washington, D.C.: The National Academies Press.
NRC. 2004. *Adaptive Management for Water Resources: Project Planning.* Washington, D.C.: The National Academies Press.
Office of Procurement Policy (OPP). 1992. Policy Letter 92-1 to the Heads of Executive Agencies and Departments, Subject: Inherently Governmental Functions. Washington, D.C.: Office of Management and Budget.
Tyler, T., and P. Degoey. 1996. "Trust in organizational authorities: The influence of motive attributions on willingness to accept decisions." In *Trust in Organizations: Frontiers of Theory and Research.* R. Kramer and T. Tyler, eds. Thousand Oaks, Calif.: Sage.
U.S. Bureau of Reclamation (USBR). 2000. *Information Management Handbook, Volume III, Drawing Management and Drafting Standards.* Denver, Colo.: U.S. Bureau of Reclamation.
USBR. 2003. *Animas-La Plata Project Construction Cost Estimates: Report to the Secretary.* Washington, D.C.: U.S. Department of the Interior.
USBR. 2005a. *Final Design Process.* Denver, Colo.: U.S. Bureau of Reclamation.
USBR. 2005b. *Reclamation Manual.* Available at http://www.usbr.gov/recman/. Accessed August 18, 2005.
USBR. 2005c. *Technical Service Center Project Management Guidelines.* Denver, Colo.: U.S. Bureau of Reclamation.
USBR. 2005d. *Technical Service Center Operating Guidelines.* Denver, Colo.: U.S. Bureau of Reclamation.

4

Workforce and Human Resources

INTRODUCTION

Reclamation is a highly professional engineering organization that historically has accomplished heroic feats of water management in the 17 western states. The days of these feats are, by most accounts, over, and Reclamation is in a new era. This new era is marked by two new tasks: (1) the operation, maintenance, and rehabilitation of existing structures and systems and (2) the creation and nurturing of brokered agreements among a variety of players affected by the management of water resources. The two tasks are interdependent, with operation, maintenance, and rehabilitation of existing structures often requiring the creation and nurturing of brokered agreements among a variety of different players. The growing need to include a broader spectrum of stakeholders, particularly groups that represent environmental issues and American Indian water rights, considerably affects how the bureau carries out its second task and the skills it requires for this.

The bureau, like other engineering organizations (e.g., USACE and TVA), faces an impending change in the workforce due to the large number of engineers and other staff who will soon retire. This change is exacerbated in Reclamation by the loss of many engineers who took early retirement in the mid-1990s. The small number of engineers graduating from engineering schools intensifies the challenge of maintaining an effective workforce.

These trends in changing skill requirements and availability of qualified personnel have interrelated implications. To some extent the impending retirements create an opportunity to hire people who have different sets of skills and to change the organizational culture. However, this opportunity is offset by the loss of senior people who have developed skills that match the new tasks, the loss of institutional memory, and the scarcity of young people with engineering skills needed for Reclamation's tasks. Both the change in tasks and the need to recruit many new people will place a premium on training, offering Reclamation an opportunity to provide integrated training—training that tailors the engineering and managerial skills to suit the current tasks.

As a leader in the sustainment and management of water resources and as an organization that plans and executes much of the necessary work, Reclamation requires a highly qualified technical and tradecraft workforce. Current bureau workforce plans also acknowledge a change in necessary workforce competencies:

> Like many other government entities, Reclamation has increased opportunities for its customers to participate in the direct decision-making process concerning water, power and related resources. Managers of the future will need increased skills to manage multi-agency, multi-interest teams. The success of a project will depend on the ability to forge agreements among large numbers of participants with widely diverse backgrounds and interests. (USBR, 2003, p. V-6 in Volume I)

Incorporating these new competencies into existing practices for hiring, training, evaluating, and promoting will allow Reclamation to ensure systematically the appropriate shift in workforce capabilities and skills. As noted in Chapter 3, more outsourcing would mean a shift in core competencies, allowing Reclamation to become a smart buyer by combining technical and acquisition skills.

WORKFORCE PLANNING

Reclamation uses workforce planning as a cornerstone for the strategic management of its human capital. It completed *Workforce Plan FY 2004-2008* in September 2003. The development of this plan used a rigorous, decentralized workforce planning methodology to allocate human capital with the appropriate knowledge, skills, and abilities (KSAs). Each of the five Reclamation regions, the Denver Office, and the Commissioner's Office developed individual workforce plans, which were then incorporated into the *Workforce Plan FY 2004-2008*.

Reclamation's workforce planning follows DOI's workforce planning template, which has six parts:

- *Strategic direction.* Sets and documents assumptions, objectives, and organizational design.
- *Supply analysis.* Describes the current workforce and assesses current workload.
- *Demand analysis.* Defines the future work of the organization and describes the needed skills and knowledge.
- *Gap analysis.* Determines differences between the current workforce and the one needed to meet the future mission.
- *Solutions and implementation.* Selects actions, tools, and interventions for addressing gaps.
- *Evaluation.* Monitors and assesses the effectiveness of implemented solutions.

The following sections address Reclamation's response to its changing mission in each part of the workforce plan.

STRATEGIC DIRECTION

Reclamation functions as a decentralized organization. However, the PMTS in Denver and the five regions rely on the Commissioner's Office for policy and guidance on workforce planning. In the mid-1990s, the structure of the workforce changed dramatically in reaction to the change in mission, from water resource development to water resource management. As noted in Chapter 2, there is no universal understanding of functions to be performed, of standards to be applied, or of authority, responsibility, and accountability at each level within Reclamation. Strategic direction is Reclamation's most significant deficiency in the workforce planning process. The following is a discussion of significant issues that need to be considered.

The bureau is heavily influenced by its focus on solving engineering problems. As employees talk about their work, the difference between the way they talk about specific engineering problems and the way they talk about more amorphous problems, including multiple stakeholders with different perspectives, is unmistakable. The engineering work is clearly exciting and energizing; the people problems are not. An important aspect is the growing need to collaborate with multiple stakeholders and to take multiple perspectives into consideration. As a result, bureau employees are faced with problems that entail considerable ambiguity.

The committee has analyzed the kinds of tasks that bureau employees engage in from the standpoint of uncertainty and ambiguity, which are related but fundamentally different. "Uncertainty can be resolved by obtaining certain specifiable pieces of information" (Feldman, 1989, pp. 4-5). Uncertainty is endemic to engineering problems (Vaughan, 1995). Indeed, it is often the uncertainty that makes an engineering problem challenging, and it is part of what makes solving the problem satisfying. Creating huge new dams involves a myriad of uncertainties. The committee observed examples of how obtaining information reduces uncertainties, such as (1) figuring out the least costly and most effective way to stop seepage from Horsetooth Dam and (2) developing solutions to the problem of mitten crabs clogging the Tracy fish screen and pumping station.

Some uncertainties are more readily and immediately resolvable than others. There is often uncertainty about future effects. Thus, we do not know what impact a new dam will have on an endangered species, but we know what information we can gather to assess this impact. Note that specifying the information does not imply that the cost of obtaining the information is reasonable or even that the information is obtainable. Sometimes, instead of gathering the information directly, we estimate or predict what the information is likely to be.

Ambiguity, on the other hand, is "the state of having many ways of thinking about the same circumstances or phenomena" (Feldman, 1989, p. 5). Specific pieces of information will not resolve ambiguity. Indeed, though gathering information is often necessary in the face of ambiguity, more information often increases the ambiguity rather than decreasing it. The appropriate balance between environmental concerns and economic concerns is an ambiguous issue. There is no right answer. Answers are matters of interpretation and will vary depending on one's perspective.

Some of the uncertainties in solving engineering problems are fundamentally irresolvable (Vaughan, 1995). The appropriate balance between cost and safety, for instance, is often sought in engineering projects and is certainly an important issue in building and repairing dams. Vaughan describes engineering work as "guided by a system of flexible rules tailored and retailored to suit an evolving knowledge base" (Vaughn, 1995, p. 203). Ambiguity increases exponentially, however, when different knowledge bases as well as different values are involved. Thus, multiple stakeholders agreeing on trade-offs involves much more ambiguity than figuring out how to implement the trade-offs that are agreed upon.

The tasks that the bureau engages in can be roughly divided into engineering tasks and resource management tasks, where the former involve less ambiguity than the latter. Table 4-1 summarizes the difference between engineering tasks and resource management tasks.

In Reclamation, two factors are influencing the changes in workforce requirements. One is that an increasing amount of the bureau's work involves forging agreements between multiple stakeholders. The other is that the increasing proportion of work that involves uncertainty also requires many stakeholders to agree in order to take action and evaluate outcomes. To be effective in the face of these changes, the bureau needs to

WORKFORCE AND HUMAN RESOURCES

TABLE 4-1 Engineering and Resource Management Tasks

Engineering Tasks	Resource Management Tasks
Technically complex	Socially and politically complex
End points relatively well defined	Open ended
Agreed-upon methods	Appropriate methods subject to disagreement and negotiation
Relatively well-defined set of stakeholders	Fluid stakeholders
Well-defined problem	Problem definition subject to interpretation and negotiation
Standards for evaluating solutions relatively clear	Standards for evaluating solutions vary across stakeholders

accommodate them by the ways it organizes work, recruits workers, and structures incentives for employees.

The changes in workforce requirements have implications for both leadership and management. Leadership is concerned with setting direction and defining the organizational culture and its mission. Management is concerned with what needs to be done to accomplish the organization's mission.

Implications for Leadership

Leadership involves actions to influence what bureau personnel think the organization is supposed to do (i.e., the organization's vision), as well as how it is perceived by others and how it perceives itself in relation to others (i.e., its image). Whether the bureau solves engineering problems or leads processes of collaboration makes a big difference for both the vision and the image.

The committee's interviews with Reclamation personnel indicated that substantial institutional memory has been lost in recent years through retirement. Two things are lost: knowledge of specific engineering projects and knowledge of stakeholders. The latter knowledge is not just knowledge of who the stakeholders are but an understanding of their perspectives and, even more important, the relationships of trust that have built up over years of interaction.

Reclamation's roles are evolving. At the same time as an increasing proportion of work is essentially negotiation and communication, there is still a role for Reclamation to play in more traditional engineering projects, such as repairing aging infrastructure and dams. Reclamation needs to have both kinds of skills, but it is not clear whether the vision is that these skills will be integrated within individuals or within the organization. If they are to be integrated within individuals, hiring will have to reflect this goal. If they are to be integrated within the organization, efforts will have to be made to value both kinds of skills and enable groups with different skills to communicate with one another.

How the bureau perceives itself in relation to other stakeholders is another aspect of leadership. Although the committee's discussions with Reclamation employees revealed generally high morale, some employees expressed a sense of victimization and resignation more than a sense of empowerment. They seemed to feel they had an impossible task and would be held responsible for not accomplishing it. This relates directly to the change in the kinds of tasks that need to be accomplished in Reclamation and the difficulty in recognizing the tasks and acknowledging how they are going.

Communication from the leadership needs to cover a wide range of activities and is critical to the successful implementation of all of Reclamation's existing programs. For instance, having a communications plan for ongoing A-76 competitive outsourcing efforts is necessary to reinforce strategic direction from the Commissioner's Office and to allay anxiety among the staff due to a lack of information. The need for more structured communication to educate new management staff will continue to increase as attrition through retirements of senior personnel peaks in 2009. Adequate funding of communications programs in both the Commissioner's Office and Human Resources (HR) will be critical for conveying strategic direction as well as for the effective use of existing HR programs.

Implications for Management

There are two broad models for taking action in the public arena in the face of uncertainty and ambiguity. One is primarily oriented to enabling action in the face of uncertainty; the other to enabling action in the face of ambiguity.

Adaptive Management

Adaptive management is a model oriented to enabling action in the face of uncertainty. It has been used in a variety of fields but is most common in the field of environmental policy (Hollings, 1978). This model promotes the use of quasi-experiments as part of the policy process (Jacobs and Westcoat, 2002). It involves taking action while there is still considerable uncertainty about outcomes but designing the action so that it can be monitored and adjusted as its effects become more clearly understood. "Management policies are designed to be flexible and are subject to adjustment in an iterative social learning process (Lee, 1999)" (NRC, 2004, p. 20). While there is no exact formula for adaptive management, its elements generally include these (NRC, 2004, pp. 24-27):

- Management objectives that are regularly revisited and accordingly revised.
- A model(s) of the system being managed.
- A range of management choices.
- Monitoring and evaluation of outcomes.
- A mechanism for incorporating learning into future decisions.
- A collaborative structure for stakeholder participation and learning.

Inclusive Management

Inclusive management is primarily oriented to enabling action in the face of ambiguity. This approach is defined as (1) a continuous iterative process that helps to create an inclusive community of participation and (2) a collective process in which a wide range of perspectives plays a role in policy making and implementation (Feldman and Khademian, 2000, 2005). The model is based on understanding the importance of combining multiple perspectives in problem-solving efforts. A rich literature explores the potential of public management directly engaged with the public to enhance the quality of public programs and strengthen democratic practices (Roberts, 2004). Consistent with this premise, managers of inclusion endeavor to facilitate the participation of a broad array of stakeholders, to put all possible options on the table, and to give stakeholders an opportunity to come to common agreement on issues of ambiguity. In the committee's discussions with the TVA, USACE, and DWR, all three organizations provided support for this or a similar approach. The TVA provided a textbook case of this kind of management in its development of priorities for revised reservoir operating plans. USACE and DWR provided more abstract support for the necessity of an inclusive approach to water resources management.

SUPPLY ANALYSIS

Workforce Plan FY 2004-2008 provides an excellent method for analyzing the supply of human capital (USBR, 2003). Over the past decade, Reclamation's workforce has been reduced by more than 25 percent, with the most significant portion of the reduction having taken place during

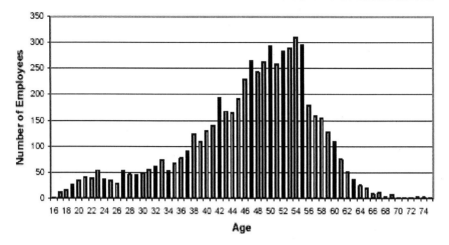

FIGURE 4-1 Distribution of Reclamation personnel by age, end of FY 2002. SOURCE: USBR, 2003.

the 1994 reorganization. The reduction has been in response to the change of mission, from water resources development to water resources management.

Reclamation's current workforce of approximately 5,900 is primarily male (65 percent), middle-aged (average age 47), white (84 percent), college-educated, professional/technical, full-time (95 percent), and permanent (93 percent). These demographics reflect those of similar private industry organizations, except that Reclamation's average age is higher (see Figure 4-1).

Reclamation anticipates approximately 7 percent annual attrition in the permanent workforce (just under half is due to retirement) and 100 percent annual attrition in temporary workforce; it also expects workforce size and occupational profile to remain relatively stable over the next 5 years (no major restructuring is currently planned). Every year approximately 400 permanent and 400 temporary employees must be hired, primarily to address attrition.

A large portion of Reclamation's workforce is nearing retirement. The workforce will be further challenged because recruiting has been heavily targeted to new graduates and there are few employees in a position to take over the responsibilities of senior personnel as they retire. The bureau now needs to keep senior expertise long enough to allow the transfer of knowledge, with one way of doing this being to use experienced consultants. A policy making it clear that retention of institutional knowledge is crucial for Reclamation would facilitate the use of retirees for mentoring and training of young personnel and provide a secondary benefit of supplementing the workforce when necessary.

DEMAND ANALYSIS

The five regional offices and PMTS in Denver have based their analysis of future workloads on anticipated future budgets. The Commissioner's Office has described out-year budgets as flat or declining. The regions and the Denver offices believe they are adequately staffed given the expectation of flat or declining budgets and limited change to Reclamation's current mission. Thus, Reclamation is predicting little change to its workforce needs in terms of either quantity or occupational profile.

This demand analysis is deficient in a variety of ways, three of which are discussed in this section. First, the competencies required to forge agreements among large numbers of participants with very diverse backgrounds and interests have not been systematically identified in the demand analysis. Second, the call for increased outsourcing of nongovernmental functions, such as facility O&M functions and noncritical engineering and science functions, to comply with the President's management objectives needs to be considered. Third, the shift from new construction to O&M tasks has not been fully incorporated into the structure of the workforce.

Forging Agreements

Reclamation employees are engaged in many efforts that require technical expertise in forging agreements. For example, creating water management plans for multiple integrated facilities in a watershed is an activity that takes place in all regions. Another example is the Lower Colorado River Multi-Species Conservation Program (MSCP), referred to in Chapter 3. MSCP is a coordinated, comprehensive, long-term, multiagency effort to conserve and recover endangered species and to protect and maintain wildlife habitat on the lower Colorado River. This program involves a 35-member steering committee, three states, and 40 customer representatives. For Reclamation to manage water resources effectively, it needs to immediately define the necessary expertise and draw up a plan to cultivate a highly collaborative staff who can troubleshoot problems, provide adequate direction to contractors, and manage risks associated with critical infrastructure and resources.

Outsourcing

Workforce Plan FY 2004-2008 does not contemplate any major shifts in workforce. Reclamation, however, is required to assess positions according to criteria established in OMB Circular A-76. As noted in Chapter 3, a strict reading of A-76 would likely find only a limited number of inherently governmental functions being performed by Reclamation's TSC and regional staff and would probably alter the demand analysis accordingly.

Project Management

While Reclamation will continue to have a sizable construction program over the next several years, clearly the mix of projects is changing. The era of megaprojects like the Hoover, Grand Coulee, and Glen Canyon dams is over, and the trend in new construction projects is to more but smaller projects for water storage and distribution systems. In addition, improvements in technology offer opportunities to increase efficiency through replacement or modification of existing equipment. This work is now done partly with in-house forces and partly by contract, depending on personnel availability and capability. As experienced craft personnel retire, the proportion of work contracted out will undoubtedly increase.

While the fundamental technical skills and procedures for managing O&M projects are the same as those for new construction projects, better social and political skills are required to advance multiagency, multi-interest projects. The owner's role in planning, design, and quality assurance/quality control (QA/QC) functions requires some different expertise, which is, however, already resident in Reclamation. Accordingly, the need for personnel with planning, design, construction management, and project management skills will continue indefinitely despite the notion that Reclamation construction is over. Because of increased outsourcing, successful completion of Reclamation's mission will also require the integration of acquisition skills with technical, managerial, and collaboration skills.

GAP ANALYSIS

Gap analysis is a determination of the difference between the number of employees currently on board and the number that are needed. When these two are correctly specified, gap analysis is straightforward. Reclamation's gap analysis identifies the following trends:

- The workforce is expected to remain relatively constant in both size and profile.

- The annual attrition rate is anticipated to be about 7 percent of the permanent workforce.
- The annual attrition of the temporary workforce is anticipated to be 100 percent.

Accordingly, there is an average annual workforce gap of approximately 400 permanent employees and 400 temporary employees.

Deficiencies in the demand analysis make gap analysis problematic. The problems are due to a failure to accommodate the change in needed competences that comes from (1) a likely increase in outsourcing and (2) the continuing shift of mission from water resource development to water resource management. These changes in needed competencies will require a change in hiring, training, evaluation, and promotion.

Engineering and resource management KSAs need to be integrated. Integration can occur in a number of ways:

- Within individuals
- Across individuals, within units
  —Functionally organized
  —Hierarchically organized
- Across units
  —Functionally organized
  —Hierarchically organized

Integration within individuals means finding people with both engineering and resource management skills. The following strategies would be useful:

- Identify specific KSAs appropriate for resource management tasks (e.g., conflict resolution, negotiation, knowledge of water rights legislation, and environmental background) and recruit engineers with these KSAs.
- Work with engineering programs to develop appropriate curricula that prepare engineers for resource management tasks.
- Provide in-house training in resource management KSAs.

The alternative to recruiting or developing personnel who have all the necessary KSAs is to develop teams whose combined KSAs fit the bill. A team approach requires individual efforts to be integrated within units or across units, which implies a greater reliance on collaborative processes. The likelihood of successful collaboration is enhanced by techniques such as the development of boundary objects that create opportunities to understand different perspectives (Feldman and Khademian,

2005). Boundary objects can be artifacts, documents, or vocabulary that are shared but interpreted differently by the different communities. The acknowledgement and discussion of these differences enables a shared understanding. An "effective boundary object facilitates a process where individuals can jointly transform their knowledge" (Carlile, 2002). Research has shown how boundary objects enable people with different perspectives to come to know something in common (Carlile, 2002).

SOLUTIONS AND IMPLEMENTATION

Reclamation has several human resource initiatives under way to meet anticipated recruitment and retention goals. This section reviews the tools and techniques of those initiatives and discusses how they could be used to even better effect.

Hiring

The Department of the Interior, including Reclamation, is currently taking action to streamline and enhance its recruitment process by centralizing legal and data management and candidate tracking for the recruitment process. Reclamation has recognized centralized candidate tracking as a key to improving the efficiency of its recruiting process. Additionally, Reclamation is evaluating programs such as QuickHire, a Web-based automated recruiting system, to speed the recruiting process and to push hiring authority to the lowest appropriate level. The actual recruiting of personnel is generally decentralized, with each of the five regions maintaining its own recruiter. Each region and the service organizations in Denver are responsible for balancing their own staff and workload. An ad hoc recruitment task force with representatives from each region, Denver, and the Commissioner's Office has been assembled to act on critical/difficult hires Reclamation-wide.

Reclamation has several programs at its disposal both to make it more visible to potential candidates and to keep it competitive within the market when filling critical positions. The Student Career Employment Program and its companion, the Student Temporary Employment Program, bring college students to the worksite for training, exposing potential recruits to Reclamation and at the same time allowing Reclamation to evaluate them. Reclamation has actively used the programs and reported good results. The regions told the committee that they would like to see more aggressive use of the Federal Career Intern Program. Reclamation has yet to outline the types of positions and responsibilities it envisions for this 2-year internship program.

Recruiting midcareer professionals is another promising avenue for acquiring technical as well as managerial competencies. The federal government can offer a competitive salary and is seen by personnel with several years of technical and management experience in the private sector as an attractive employer.

Reclamation uses recruitment bonuses, relocation bonuses, and student loan repayment programs to remain competitive in the market when filling critical positions. These inducements currently require Commissioner's Office approval. Regional staff said that it is too difficult and time-consuming to implement these programs and that they may be constrained by a lack of funds.

All of these tools work well to ensure that people are hired, but it is not clear that they are being used systematically to bring Reclamation the new competencies necessitated by the change from water resources development to water resources management. The bureau needs to be more disciplined in defining the required competencies and to include them in the profiling and screening processes. The committee notes that in *Workforce Plan FY 2004-2008*, only one region (Mid-Pacific) specifically related competencies to job categories. Without such efforts, it is difficult to tell where new competencies are required and to track whether the need for new competencies is being assessed on a regular basis.

A structured interviewing approach might also allow newly identified competencies to be sought out in the recruitment process. A structured process would provide an organized and comprehensive system to identify critical competencies for particular positions, evaluate a candidate's past performance to predict future performance, teach interviewers effective interviewing techniques, and provide for organized data exchange between multiple interviewers.

Training and Mentoring

Reclamation has traditionally been an engineering- and science-driven organization. As such, training has been heavily focused on basic technical competencies. The success of Reclamation's mission to manage water resources will more and more depend on the bureau's ability to solve problems through consensus, requiring an increased emphasis on training and the retention of staff with collaborative competencies at all levels of the organization. Additionally, as the bureau more directly attempts to determine the right mix between contractor and in-house support, it should also ensure that in-house staff has the overall technical expertise to be able to monitor contractor performance effectively. Reclamation has many managers who require extensive training to perform the contracting officer's technical representative (COTR) function, and it should reassess its existing career development programs to make sure that they provide this training. Moreover it should explicitly recognize the important strategic role of the COTR in accomplishing the mission.

The type of training will depend on how the engineering and resource management competencies are to be integrated. One kind of training will provide engineers with collaborative skills; another will provide teams of people with the skills to work together effectively.

On a limited basis, Reclamation uses individual development plans (IDPs) to identify training needs for specific individuals. Additionally, IDPs improve employee retention and morale by engaging supervisors and employees in a mentoring and planning process that promotes professional development consistent with the bureau's strategic direction. IDPs become the communication link that synchronizes organizational goals and needs with employee capabilities. Reclamation should mandate the use of IDPs to improve overall communication, to allocate resources, to take better advantage of personnel KSAs, and to plan for training.

Reclamation has recognized the graying of its workforce, and its current workforce plans incorporate ways to maintain and transfer specialized knowledge and skills to younger members of the workforce. Reclamation has the good fortune of having a skilled and dedicated senior workforce. Many of its employees are working beyond the time they are eligible to retire. Reclamation has been successfully using retention bonuses to keep the services of key senior personnel who are eligible for retirement. As an alternative, Reclamation employees in jeopardy of reducing their retirement benefits by delaying retirement have entered into postretirement contracts with Reclamation. Taking advantage of this situation requires coordination between Human Resources and Contracting to accommodate the potential for increased outsourcing to retirees. Both approaches are allowing Reclamation additional time to hire and train new personnel as incumbents retire.

The committee learned that Reclamation has had a program for rotating the assignments of new hires, but that the program has been largely abandoned because of cost constraints. Such a rotation program can provide a broad range of experience and help to develop collaborative competencies. The committee believes that Reclamation should restart these rotations and that the assignments should entail a variety of technical experiences, including construction, and offer opportunities to engage in making policy and forging agreements. The program can be used as a tool for recruiting, training, and mentoring, as well as for enhancing retention.

In the past, Reclamation has been a leading member of the International Commission on Irrigation and Drainage and the International Commission on Large Dams and very a strong supporter of both. International activities have been considerably scaled back and currently consist of technical assistance programs in Iraq (river basin modeling) and Israel (dam safety) and hosting international workshops on integrated water resources management, modern methods in canal operation and control, and dam safety operation and maintenance. The international unit also assists the U.S. Virgin Islands with environmental assessment. International activities not only enhance Reclamation's prestige but are also a valuable tool for recruitment, training, mentoring, and retention and should be considered for future funding.

Employee Motivation

Employee motivation is an important part of managing any organization. The challenge presented by the shift in Reclamation's tasks is how to motivate employees who gain satisfaction from creating things and solving technical problems to also gain satisfaction from negotiating complex social arrangements.

A strategy of "small wins," described below, seems appropriate for managing the complex social tasks that Reclamation is called on to perform. Karl Weick (1984) argued that shifting attention from outcomes to inputs may be a useful way to bring out the best in people's problem-solving abilities. The psychological research described by Weick shows that there is a U-shaped relationship between the physiological states that accompany stress and anxiety (arousal) and those associated with performance efficiency and that the optimal level of arousal varies inversely with the difficulty of the task—that is, a very difficult task calls for very low stress. When people become too stressed, coping responses become more primitive (Staw et al., 1981, summarized in Weick, 1984). People tend to process fewer cues and revert to earlier, often less finely tuned ways of coping. Breaking problems down into smaller, more manageable chunks enables people to attend to the problem in ways that enhance their problem-solving abilities. Weick argued that this will not only bring out the best in the people working on the problem but will also lead to "wins" that can be built upon.

This small win strategy seems very much applicable to the issues confronting Reclamation's employees. Confronted with the complex problems currently facing the bureau, any reasonable person would throw up his or her hands. Responsibility for an overall outcome appears beyond reach for a single individual. Responsibility for some features of an overall process, however, might not only be manageable but also interesting and fun. Features of the process might include engaging in a series of stakeholder analyses (Bryson, 2004) or facilitating opportunities for stakeholders to communicate with one another (Crosby and Bryson, 1992).

Another aspect of such small wins is that they provide opportunities for celebrating successes. These opportunities are important for a number of reasons. First, being able to celebrate a success in the midst of a complex process gives management a chance to reward employees. Research has shown that public employees, more than employees in the private sector, are motivated by the opportunity to help and to influence public affairs (Rainey, 1997, p. 210ff). Small wins can help people see the impact they are having on complex negotiations. Extrinsic rewards, such as salary and incentive pay are also important. Again, the strategy of small wins enables managers to acknowledge gains through extrinsic rewards. Second, celebrating successes can also be helpful in creating better relations with stakeholders. Small wins let people become engaged in an effort that makes sense in the short term and that develops a strong track record for them over the long term.

Performance Evaluation and Promotion

Performance evaluations that specifically target collaborative as well as technical competencies are currently applied to Senior Executive Service staff. Similar evaluations should be used for a broader set of employees in order to encourage the development of these competencies throughout the organization.

A technically oriented individual can move up through the organization in two ways: (1) by staying on a technical track, the individual can move from being a local resource to becoming a regional or even bureauwide resource and (2) by developing more collaborative competencies, the individual can move to managerial and leadership positions.

EVALUATION

Reclamation summarized the recommendations of its workforce plan in an action item format to allow monitoring their implementation. The 13 action items described the issues and goals, identified the sponsor and team members, and provided a schedule for implementation. The committee has no information about the current status of the 13 action items; however, it applauds Reclamation for taking this approach and believes that it will help human resources management.

REFERENCES

Bryson, John M. 2004. "What to do when stakeholders matter." *Public Management Review* 6(1): 21-53.

Carlile, Paul R. 2002. "A pragmatic view of knowledge and boundaries: Boundary objects in new product development." *Organization Science* 13(4): 442-455.

Crosby, Barbara C., and John M. Bryson. 1992. *Leadership for the Common Good: Tackling Public Problems in a Shared Power World*. San Francisco, Calif. : Jossey-Bass Publishers.

Feldman, Martha S. 1989. *Order Without Design.* Stanford, Calif.: Stanford University Press.
Feldman, Martha S., and Anne M. Khademian. 2000. "Management for inclusion: Balancing control with participation." *International Public Management Journal* 3(2): 149-168.
Feldman, Martha S., and Anne M. Khademian. 2005. "Role of managers in inclusive management." Paper for National Public Management Conference.
Holling, C.S., ed. 1978. *Adaptive Environmental Assessment and Management.* New York, N.Y.: John Wiley and Sons.
Jacobs, J.W., and J.L. Wescoat, Jr. 2002. "Managing river resources: Lessons from Glen Canyon Dam." *Environment* 44(2): 21-31.
Lee, K.N. 1999. "Appraising adaptive management." *Conservation Ecology* 3(2): 3.
National Research Council (NRC). 2004. *Adaptive Management for Water Resources: Project Planning.* Washington, D.C.: The National Academies Press.
Rainey, Hal G. 1997. *Understanding and Managing Public Organizations,* 2nd edition. San Francisco, Calif.: Jossey-Bass Publishers.
Roberts, Nancy. 2004. "Public deliberation in an age of direct citizen participation." *American Review of Public Administration* 34(4): 315-353.
Staw, Barry M., Lloyd E. Sandelands, and Jane E. Dutton. 1981. "Threat-rigidity effects in organizational behavior: A multi-level analysis." *Administrative Science Quarterly* 26: 501-524.
U.S. Bureau of Reclamation (USBR). 2003. *Workforce Plan FY 2004-2008: Volume I and Volume II.* Washington, D.C.: Department of the Interior.
Vaughan, Diane. 1995. *The Challenger Launch Decision: Risky Technology, Culture and Deviance at NASA.* Chicago, Ill.: The University of Chicago Press.
Weick, Karl E. 1984. "Small wins: Redefining the scale of social problems." *American Psychologist* 39(1): 40-49.

5

Alternative Scenarios for Future Infrastructure Management

INTRODUCTON

The committee considered a broad range of alternative scenarios as it contemplated Reclamation's future responsibility and its organization for construction and infrastructure management. They ranged from scenarios that dramatically expanded Reclamation's mission to scenarios that eliminated the bureau and redistributed its responsibilities to other existing agencies. Because the alternatives at the extreme ends of the spectrum were deemed to be improbable, they were not analyzed further. The committee agreed on three scenarios it believes will provoke productive thought and be of maximum assistance to Reclamation and the Department of the Interior. They are considered feasible, consistent with national trends and stakeholder interests, and responsive to the trends as identified and described in this report. These scenarios do not predict future requirements nor are they recommendations of the committee—they are only intended to stimulate discussion.

Reclamation has recognized and taken steps to adapt its tasks as it changes from water resource development to water resource management. This change has turned Reclamation from a construction and capital-oriented organization into an operations and maintenance organization that requires determining the appropriate balance and borders between centralized policy and decentralized operations. The following scenarios describe how these trends might affect the way Reclamation constructs and maintains facilities to deliver power and water.

ALTERNATIVE SCENARIOS FOR FUTURE INFRASTRUCTURE MANAGEMENT

The trends discussed in the previous chapters that had particular influence on the development of the scenarios are these:

- The O&M workload is growing and is expected to continue to grow.
- The major construction workload is diminishing, and the source and kind of new construction activity are uncertain.
- The construction workload will be driven by dam safety considerations, environmental mitigation and restoration projects, small projects incident to maintenance and operations, larger rehabilitation, repair, and modernization projects, and new construction to satisfy American Indian water rights.
- Current federal policy, embraced by officials of all political parties, will continue to encourage the transfer of field execution activities, to the extent possible, from government employees to contractors.
- In response to their requests, water users will be increasingly responsible for transferred works, but with Reclamation guidance and technical assistance. Water districts and other users will be free to accomplish more of the design and construction incident to O&M.
- Water users will be required to provide an increasing proportion of O&M financing, and as facilities age, rehabilitation and repair will become larger components of the budget.

The current line organization flowing from the commissioner to the regional director to the area manager appears simple, efficient, and responsive to mission demands. This organization is considered a given in all of the scenarios. The provision of technical and administrative services from a central organization is also responsive; however, the size of the central service organization relative to that of the line organization service units is likely to change along with their roles. Though the basic organization remains intact, the number of personnel at each level and the knowledge, skills, and abilities to complete the assigned tasks vary dramatically from scenario to scenario.

Scenarios 2 and 3 could occur concurrently with Scenario 1. For a single project, Scenarios 2 and 3 are mutually exclusive, but they could be occurring concurrently on different projects.

SCENARIO 1:
CENTRALLY LOCATED PROJECT
MANAGEMENT ORGANIZATION

Construction projects other than minor projects that are undertaken by area or regional offices are executed by a centrally located construction project management organization. Minor projects are defined as the commissioner may direct according to cost (e.g., less than $5 million) and/or complexity and risk. The regional office remains the owner of the project, but this scenario is based on a reduction in the number of major projects, making it impractical to maintain the necessary competencies at the regional level. As the owner, the region plays a significant role in early planning activities to define the project scope, but control is shifted to a central organization as the project progresses. This scenario also assumes that outsourcing of design services will increase to the point where it is the predominant means of implementing projects. The central organization provides project management services, thus overseeing design and construction activities. Unit personnel, while based at a central location, are deployed as needed to field locations to execute the construction task. Upon completion, the construction unit transfers ownership responsibilities for O&M to the assigned organization.

Reclamation recognizes the growing predominance of O&M tasks and responsibilities and the diminishing importance of but continuing need for a viable construction capability. There is an obligation to maintain a broadly based field organization for stakeholder interaction and support and for water and power contract oversight and administration. The existing organization of regional and area offices is well suited to the execution of O&M tasks, including minor construction projects.

Scenario 1 implies the following organizational characteristics:

- Project management and construction expertise for major construction projects will be concentrated in a centrally located unit and largely stripped from the existing organization.
- The central project management unit will include personnel with skills and qualifications to serve as contracting officers; to oversee design provided by the regional offices, by TSC, or by contract; to supervise contract or construction activities in the field; and to ensure integration of user needs as determined by line organization managers. The unit will perform all of the functions of a smart buyer—that is, it will ensure proper project scoping; selection of an appropriate project execution strategy and contractors; and administration of the contracts on behalf of Reclamation and will conduct quality assurance activities.
- The central project management unit, consisting of a more or less fixed number of highly qualified specialists, will continue to charge the costs of services to projects but may also require nonproject funding support to maintain its core competencies. The unit will be augmented by contract consultants during periods of heavy workload.

SCENARIO 2:
OUTSOURCED OPERATIONS AND MAINTENANCE

Outsourcing of essentially nongovernmental functions increases to the point where Reclamation accomplishes all of its field O&M tasks by contract except those determined to be inherently the government's responsibility. The O&M for major hydroelectric plants and dams that pose the most significant risks is likely to continue to be a Reclamation function, but with increasing support services by contractors. The bureau retains a line regional and area structure to execute and administer contracts, to interact with stakeholders and water and power contract partners, and to discharge governmental responsibilities of ownership.

This scenario is consistent with current government-wide goals of increasing the outsourcing of nongovernmental functions. It opens up opportunities for local entities to perform many O&M functions on their own projects. Having motivated providers in charge would presumably result in reduced costs. It allows greater stakeholder involvement in ongoing operations while reducing the need for Reclamation employee involvement.

Scenario 2 implies the following organizational characteristics:

- Only Reclamation's nongovernmental functions may be outsourced. Reclamation can compete with private organizations for O&M contracts, but the competitive sourcing process makes it difficult for government-provided operations to be reinstated after they have been shifted to contractors. Water district partners are free to choose their preferred method of executing the program elements for which they are responsible.
- Reclamation staff will learn to be smart buyers, and procurement and contract oversight and administration specialists will be trained.
- More emphasis will be placed on developing standards and guidelines necessary to facilitate contract scoping and identify mandatory procedures.

SCENARIO 3:
FEDERAL FUNDING AND LOCAL EXECUTION

This scenario further reduces Reclamation's direct involvement in the management of assets. Under it, Reclamation administers its O&M program by distributing federal funds to the irrigation and power users in response to project needs. The users are held responsible for project O&M in conformity with Reclamation standards and guidelines, which are designed to ensure maximum flexibility within the bounds of essential public health and safety interests.

Reclamation retains responsibility for essential governmental policy and oversight, necessitating close and continuing communication and interaction between the recipients of funds and Reclamation officials. The emphasis is on Reclamation exercising an oversight function to assure that its standards and guidelines are respected by water and power users.

Scenario 3 implies the following organizational characteristics:

- Reclamation personnel skills will change from direct involvement in task execution to administration of a federal funds program in support of what had traditionally been Reclamation responsibilities. Reclamation's efforts will include needs validation, priority determination, defense of appropriations requests, and program oversight to assure faithful application of resources.
- In spite of fundamental program administrative changes, Reclamation will retain responsibility for stakeholder interaction and communications.

CONCLUSION

The scenarios described above are not predictions about the future. They are based on current trends which are taken to a logical, but not necessarily probable, conclusion. They are not the only scenarios that could have been developed. These three scenarios are all based on Reclamation having an organizational structure that is the same as or very close to its current structure. Other scenarios could be based on other organizational forms (e.g., regional offices that operate as independent organizations or a strong central administration without regional offices) and could be applied to the same basic concepts with different results.

Irrespective of which models are implemented in the future, Reclamation will continue to have responsibility for program and project planning as stewards of water and land resources in the West. This responsibility will require continuing assessment of the existing water management infrastructure, new physical and operational systems, and the need to evaluate and prioritize among all of them. A recent review of USACE water resources planning (NRC, 2004) recommended a portfolio planning process that considers issues such as the operational benefits that may be realized when investment in a new project results in increased value of the water infrastructure. A number of principles are stated that, if followed, could guide the planning process. Adopting a similar approach could prove beneficial in any of the three scenarios.

The committee considers these scenarios as a starting point. This report would not have been possible without extensive input from Reclamation managers, but much more is needed to make scenario planning an effective management tool for the bureau. More extensive and active participation of Reclamation personnel in scenario development will help managers break away from current assumptions, disclosing the possible threats and opportunities that may have been missed. Active scenario planning can also disclose possible implications of current events and policy decisions and help to create boundary objects to help bring together divergent ideas and opinions in the bureau.

The three scenarios presented here are just a starting point insofar as additional input from Reclamation managers is needed to determine what the bureau will need to do to succeed in each of these possible futures. Exponential increases in technology are hastening the rate of change in management of government agencies. Reclamation, like other agencies, needs to be able to recognize future requirements so that it can be prepared to meet them. The continued involvement of Reclamation managers in scenario planning can follow up on what this report has begun by identifying emerging patterns of factors that shape the bureau's mission, extrapolating the past into the future, identifying cycles and patterns that differentiate the past from the future, and using their knowledge of the goals and motivations of all stakeholders to synthesize future actions.

REFERENCE

National Research Council (NRC). 2004. *Review Procedures for Water Resources Project Planning*. Washington, D.C.: The National Academies Press.

6

Conclusions, Findings, and Recommendations

INTRODUCTION

As the study progressed it became apparent to the committee that a number of important factors, realities, and issues have major impacts on Reclamation's ability to respond quickly and effectively to the many diverse pressures and rapid changes occurring today. Equally important are the capabilities that are needed within Reclamation to deal effectively with the challenges posed by these impacts. The factors affecting the management of construction and infrastructure and the capabilities that will be needed to respond to these impacts are identified in the following sections. The findings and recommendations are based on these factors and the detailed discussions in the preceding chapters.

The history of the Bureau of Reclamation was presented to the committee in terms of six eras:[1]

- Establishment of Reclamation to the Colorado Compact, 1902-1928.
- The Depression, 1928-1941.
- World War II, 1941-1945.
- Postwar construction, 1946-1968.
- Building out after passage of the Colorado River Basin Projects Act, 1969-1988.
- Dam safety/water management, 1989-present.

---

[1] Brit Storey, Reclamation historian, "Organizational history of the Bureau of Reclamation," Presentation to the committee on February 28, 2005.

CONCLUSIONS, FINDINGS, AND RECOMMENDATIONS   95

The committee believes that Reclamation is in a new era that has been shaped by the factors impacting its mission. These diverse factors, discussed below, expand the dam safety and water management focus of the last era.

FACTORS IMPACTING THE MANAGEMENT OF CONSTRUCTION AND INFRASTRUCTURE

Although the core of Reclamation's basic mission remains much the same—to deliver water and to generate power—the way the mission is carried out is constrained by and must be responsive to several realities:

- *Environmental factors.* The environmental revolution of the last decades of the twentieth century imposed new requirements for environmental assessment, protection, and enhancement on virtually everything that the bureau does. These new requirements increase project costs and further constrain the availability of water for human uses. Consideration of the effects of a project on environmental costs and opportunities to increase sustainability must become ingrained from the outset, not simply an add-on to business as usual. Engineers and builders must be both environmental experts and water resource experts.
- *American Indian water rights and rural water needs.* America Indian water agreements and growing demands that adequate supplies of good quality water be provided to small rural communities place new demands on the regulation of river flow and storage and distribution systems.
- *Urbanization.* Land is being taken out of agricultural production in many areas of the West and being urbanized for industrial, commercial, and residential purposes. This changes the balance between irrigation and municipal and industrial (M&I) needs, which, in turn, impacts costs, treatment requirements, and the required infrastructure.
- *Increasing budget constraints.* Reclamation's budgets have been effectively shrinking for many years, even as the needs have increased. Finding new and better ways to do more with less seems to be a way of life for almost all agencies. Development of rational methods for dealing with unpredictable events when they occur and defensible techniques for prioritization of projects in a competitive environment will be major challenges.
- *Broader set of stakeholders.* Water users of all types—farmers, power distributors, consumers, homeowners, environmentalists, American Indian tribes, and virtually anybody who uses water and power in the 17 western states—are impacted by and pay in some way for what the bureau does. Many more voices want to be heard now than during the building boom of the first two-thirds of the twentieth century. As projects have

aged and O&M costs have increased, the growing financial burden on Reclamation's contract customers has increased their interest and insistence on participating in all phases of Reclamation's management processes.

- *Aging workforce.* The baby boomers will be retiring in large numbers over the next 5 to 15 years, not only from Reclamation, but also from all government agencies. This provides both challenges and opportunities for Human Resources, not the least of which will be loss of institutional memory and changes in workforce culture.
- *Aging infrastructure.* Most of Reclamation's major dams, reservoirs, hydroelectric plants, and irrigation systems are 50 years or more old. As a result, maintenance, rehabilitation, and replacement programs can be expected to form an increasing portion of Reclamation's future workload.
- *Shift from design and construction to operations and maintenance.* It is unlikely that new Hoover- and Grand Coulee-type projects will be constructed in the foreseeable future. O&M activities will form a major part of the workload. New workforce skills and interests will be needed. Outsourcing of activities that were once undertaken by Reclamation personnel is likely to grow.
- *Congressional mandates.* Political pressures, the inclusion of special mandates in new congressional legislation, and the earmarking of funds for pet projects and special interests are not new to the bureau, nor does anything in the current political climate suggest that they will ever go away.
- *Title transfer.* Transferring ownership of government-owned facilities to nonfederal agencies and the private sector, while reducing Reclamation's O&M workload, introduces budgetary and oversight issues that may necessitate new business models. Reclamation's customers vary greatly in how they feel about the desirability of accepting title to facilities.
- *Water user operation of government-owned facilities.* Reclamation has turned over and will undoubtedly continue to turn over some of its facilities to water user groups, often local water districts, for operation, maintenance, and—sometimes—rehabilitation and new construction. Equitable policies for cost sharing and recovery, distribution of user fees, oversight, and engineering, design, and construction services are needed.
- *New modes of augmenting the water supply.* In the absence of significant climate change or major technological breakthroughs, water resources will remain constant, while demand can be anticipated to increase. Droughts will have an even greater impact. It can be anticipated that the costs and environmental consequences will make constructing major new dams and storage reservoirs unlikely within the next several years. Ac-

cordingly, alternative means for meeting the water needs of the western states will need to be explored. Calls for more research and development in the areas of water conservation, water recycling, and desalination are likely to become louder and more frequent.

- *Increase in the number of small projects.* Although demand for large new projects will remain low, it is likely that demand for small water storage, irrigation, and distribution projects will increase as more and more agricultural land is transformed for municipal development. Conservancy districts and environmental restoration and enhancement projects will have special requirements where Reclamation will be a resource and have oversight responsibilities.

Findings and Recommendations

Centralized Policy and Decentralized Operations

Finding 1a. For the past decade many of Reclamation's functions have been decentralized and directed by regional office directors and area office managers. Concurrent with implementation of the decentralized organizational model, Reclamation-wide directives, known as Instructions, were withdrawn, although in some cases they continue to be used for guidance in the field. Mandatory requirements that replace the Instructions have been and continue to be developed and published as policy and directives in the *Reclamation Manual*.[2] However, some issues either have not been addressed or need additional detail. This has led to inconsistencies in understanding and implementing the functions to be performed at each level of the organization, the standards to be applied, and the authority and accountability at each level. Consistently implementing Reclamation's mission will require clear statements of policy and definitions of authority and standards.

Finding 1b. Reclamation's customers and other stakeholders want close contact with empowered Reclamation officials. They also want consistency in Reclamation policies and decisions as well as decision makers with demonstrated professional competence.

Finding 1c. Decentralization has meant that some area and project offices housing a dedicated technical group are staffed by only one or two individuals. The committee is concerned about the effectiveness of such small units and whether their technical competencies can be maintained.

---

[2]The *Reclamation Manual* is a Web-based collection of policies and directives that is continually updated and revised. Available at http://www.usbr.gov/recman/.

Recommendation 1a. To optimize the benefits of decentralization, Reclamation should promulgate policy guidance, directives, standards, and how-to documents that are consistent with the current workload. The commissioner should expedite the preparation of such documents, their distribution, and instructions for their consistent implementation.

Recommendation 1b. Reclamation's operations should remain decentralized and guided and restrained by policy but empowered at each level by authority commensurate with assigned responsibility to respond to customer and stakeholder needs. Policies, procedures, and standards should be developed centrally and implemented locally.

Recommendation 1c. The design groups in area and project offices should be consolidated in regional offices or regional technical groups to provide a critical mass that will allow optimizing technical competencies and providing efficient service. Technical skills in the area offices should focus on data collection, facility inspection and evaluation, and routine operations and maintenance.

Technical Service Center

Finding 2a. The Technical Service Center (TSC) is a large, centrally located, highly structured organization with numerous separate subunits. Many Reclamation customers and stakeholders believe that its costs are excessive, it imposes overly stringent requirements, it too often fails to complete specified work on time, and it sometimes executes projects in a manner contrary to the concept of decentralization. The size of TSC is perceived to be excessive and its organization to be inefficient.

Finding 2b. TSC's response to criticisms has been to benchmark itself against private sector architecture and engineering organizations and to adopt some private sector business practices. In an effort to remain cost competitive, TSC has developed a business plan that provides some services that are not inherently governmental.[3] A strategy of cost averaging, which blends the costs of specialized technical services and oversight with

---

[3]The basic definition of an inherently governmental function from the Office of Management and Budget Policy Letter 92-1 is as follows: "As a matter of policy, an 'inherently governmental function' is a function that is so intimately related to the public interest as to mandate performance by Government employees. These functions include those activities that require either the exercise of discretion in applying Government authority or the making of value judgments in making decisions for the Government." See Chapter 3 for a detailed discussion.

CONCLUSIONS, FINDINGS, AND RECOMMENDATIONS 99

those of other services such as collection of field data and development of construction documents, will continue to subject TSC to fire from Reclamation customers and its private sector competitors and is inconsistent with current federal outsourcing initiatives.

Finding 2c. Regional offices, area offices, water and power beneficiaries, and other stakeholders all perceive an ongoing need for a centralized, high-level center of science and engineering excellence within Reclamation. The committee believes that a thorough review and evaluation of TSC and its policies and procedures could result in a smaller, more efficient and effective TSC.

Recommendation 2a. The commissioner should undertake an in-depth review and analysis of TSC to identify the needed core technical competencies, the number of technical personnel, and how the TSC should be structured for maximum efficiency to support the high-level and complex technical needs of Reclamation and its customers. The proper size and composition of TSC are dependent on multiple factors, some interrelated:

- Forecast workload,
- Type of work anticipated,
- Definition of activities deemed to be inherently governmental,
- Situations where outsourcing may not be practical,
- Particular expertise needed to fulfill the government's oversight and liability roles,
- Personnel turnover factors that could affect the retention of expertise, and
- Needs for maintaining institutional capability.

This assessment and analysis should be undertaken by Reclamation's management and reviewed by an independent panel of experts, including stakeholders.

Recommendation 2b. The workforce should be sized to maintain the critical core competencies and technical leadership but to increase outsourcing of much of the engineering and laboratory testing work.

Recommendation 2c. Alternative means should be developed for funding the staff and operating costs necessary for maintaining core TSC competencies, thereby reducing the proportion of engineering service costs reimbursable by customers.

Reclamation Laboratory and Research Activities

Finding 3. Reclamation's laboratory and research activities came of age during the era of large dam construction in the twentieth century, when much of the needed expertise resided in the federal government and there were no laboratories capable of handling the necessary work. The needs for large materials, hydraulics, and geotechnical laboratories are much different today because the types of capabilities needed to carry out Reclamation's mission have evolved and are available from other organizations (government, university, and private). Although the need for research on environment and resource management continues to grow, the committee believes that the laboratory organization and its physical structure may be too large.

Recommendation 3a. Reclamation's Research Office and TSC laboratory facilities should be analyzed from the standpoint of which specific research and testing capabilities are required now and anticipated for the future; which of them can be found in other government organizations, academic institutions, or the private sector; which physical components should be retained; and which kinds of staffing are necessary. The assessment should also recognize that too much reliance on outside organizations can deplete an effective engineering capability that, once lost, is not likely to be regained. In making this assessment Reclamation should take into account duplication of facilities at other government agencies, opportunities for collaboration, and the possibility for broader application of numerical modeling of complex problems and systems.

Recommendation 3b. Considering that many of the same factors that influence the optimum size and configuration of the TSC engineering services also apply to the research activities and laboratories, Reclamation should consider coordinating the reviews of these two functions.

Outsourcing

Finding 4a. From its inception, Reclamation has undertaken difficult, highly technical projects with a talented and dedicated workforce of engineers and craftsmen. Reclamation's tasks have changed and the composition of its workforce has changed accordingly, but it continues to be an organization that primarily executes engineering and construction for O&M and some rehabilitation and modernization. Reclamation has been outsourcing some of its O&M functions, primarily in nontechnical areas, but could outsource more. The committee believes that many of Reclamation's activities are not what would generally be considered essentially governmental. The committee further believes that although water operations policy decisions are essentially governmental, implementation of these decisions is not and could be almost completely outsourced.

Finding 4b. Decisions on which personnel to use—area, regional, TSC, or contractors—tend to be made at the regional level and on an ad hoc basis. Decisions often hinge on the availability of federal employees to do the work. There is increasing pressure on Reclamation to allow water districts, American Indian tribes, and other customers to undertake their own planning, design, and construction management functions.

Recommendation 4. Reclamation should establish an agency-wide policy on the appropriate types and proportions of work to be outsourced to the private sector. O&M and other functions at Reclamation-owned facilities, including field data collection, drilling operations, routine engineering, and environmental studies, should be more aggressively outsourced where objectively determined to be feasible and economically beneficial.

Planning for Asset Sustainment

Finding 5a. The committee observed effective systems for planning and executing facility O&M in some regions. The 5- and 10-year plans based on conditions assessments and maintenance regimes form the core of the process. The result is an infrastructure that appears able to support Reclamation's mission for the foreseeable future.

Finding 5b. The O&M burden for an aging infrastructure will increase, and the financial resources available to Reclamation, its customers, and contractors may not be able to keep up with the increased demand. Some water customers already find full payment of O&M activities difficult, and major repairs and modernization needs, if included in the O&M budget, impose an even greater financial burden that cannot be met under the current repayment requirements. Long-term sustainment will require more innovation and greater efficiency in order to get the job done.

Finding 5c. The committee observed extensive efforts and success in benchmarking Reclamation's hydropower activities; however, there appears to be little effort to benchmark the O&M of water distribution facilities. The committee believes that benchmarking can help improve the efficiency of Reclamation's water management and distribution activities as well as those of the water contractors responsible for transferred works.

Recommendation 5a. Because effective planning is the key to effective operations and maintenance, Reclamation should identify, adapt, and adopt good practices for inspections and O&M plan development for bureauwide use. Those now in use by the Lower Colorado and Pacific Northwest regions would be good models.

Recommendation 5b. Reclamation should formulate comprehensive O&M plans as the basis for financial management and the development of fair and affordable repayment schedules. Reclamation should assist its customers in their efforts to address economic constraints by adapting repayment requirements that ease borrowing requirements and extend repayment periods.

Recommendation 5c. Benchmarking of water distribution and irrigation activities by Reclamation and its contractors should be a regular part of their ongoing activities.

Project Management

Finding 6a. Reclamation does not have a structured project management process to administer planning, design, and construction activities from inception through completion of construction and the beginning of O&M. Projects are developed in three phases: (1) planning (including appraisal, feasibility, and preliminary design studies), (2) construction (including final design), and (3) O&M, with each phase having a different management process.

Finding 6b. The *Reclamation Manual* includes a set of directives for managing projects, but it is incomplete, and there is insufficient oversight of its implementation. Central oversight of some projects is being developed in the Design, Estimating, and Construction Office, but policies and procedures have not yet been completed.

Finding 6c. Reclamation needs to recognize project management as a discipline requiring specific knowledge, skills, and abilities and to require project management training and certification for its personnel who are responsible for project performance. The committee observed the appointment of activity managers in the Pacific Northwest region who were responsible for communications and coordination among project participants for all phases of the project. These activity managers appeared to be beneficial for the execution of projects, but the committee believes that a project manager with responsibility and authority to oversee projects from inception to completion could be even more effective.

Finding 6d. Reclamation has long-standing experience and expertise in planning, designing, and constructing water management and hydroelectric facilities, yet recurring problems are affecting the agency's credibility for estimating project costs. The cost estimating problems associated with the Animas–La Plata Project are a notable example. This project was submitted for appropriations with an incomplete estimate and became a serious problem for Reclamation. Comprehensive directives on the cost estimating process have been drafted but have not yet been published. These directives require that a feasibility estimate must be completed before a project is submitted for appropriations.

Recommendation 6a. Reclamation should establish a comprehensive and structured project management process for managing projects and stakeholder engagement from inception through completion and the beginning of O&M.

Recommendation 6b. Reclamation should develop a comprehensive set of directives on project management and stakeholder engagement that is similar to TSC directives for agency-wide use.

Recommendation 6c. Reclamation should establish a structured project review process to ensure effective oversight from inception through completion of construction and the beginning of O&M. The level of review should be consistent with the cost and inherent risk of the project and include the direct participation of the commissioner or his or her designated representative in oversight of large or high-risk projects. The criteria for review procedures, processes, documentation, and expectations at each phase of the project need to be developed and applied to all projects, including those approved at the regional level.

Recommendation 6d. A training program that incorporates current project management and stakeholder engagement tools should be developed and required for all personnel with project management responsibilities. In addition, project managers should have professional certification and experience commensurate with their responsibilities.

Recommendation 6e. Reclamation should give high priority to completing and publishing cost estimating directives and resist pressures to submit projects to Congress with incomplete project planning. Cost estimates that are submitted should be supported by documents for design concept and planning, environmental assessment, and design development that are sufficiently complete to support the estimates. Reclamation should develop a consistent process for evaluating project planning and the accuracy of cost estimates.

Acquisition and Contracting

Finding 7. Different Reclamation regions employ different contracting approaches and use a variety of contracting vehicles to meet their acquisition needs. These range from indefinite delivery/indefinite quantity (IDIQ) contracts with multiple vendors to reverse auction or performance-based contracting techniques to achieve more cost-effective results. In addition, regions are employing innovative approaches for maintaining stakeholder involvement in the contracting process.

Recommendation 7. Reclamation should establish a procedure and a central repository for examples of contracting approaches and templates that could be applied to the wide array of contracts in use. This repository should be continually maintained and upgraded to allow staff to access lessons learned from use of these instruments.

Relationships with Sponsors and Stakeholders

Finding 8. The committee believes that the key to effective relationships between Reclamation and its sponsors and stakeholders is open communication and an inclusive process for developing measures of success. In addition, the more transparent and consistent the processes used by Reclamation, the easier it will be to obtain buy-in from sponsors and stakeholders. The Lower Colorado Dams Office's interactions with its coordinating committee of sponsors illustrate the beneficial effects of these factors and their contribution to successful operation of the project.

Recommendation 8. Making information readily available about processes and practices, both in general and for specific projects and activities, should be a Reclamation priority. Successful practices, such as those used in the Lower Colorado Dams Office, should be analyzed and the lessons learned should be transferred, where practical, throughout the bureau.

CAPABILITIES FOR THE MANAGEMENT OF CONSTRUCTION AND INFRASTRUCTURE

Dealing with the challenges identified in the preceding section will necessitate a workforce with special skills and a mindset that can look at old problems in new ways and attack new problems effectively. Committee members were most favorably impressed by the high morale, enthusiasm, optimism, loyalty, and dedication to mission of the Reclamation personnel they met during this study. Building on these strengths, even as the aging workforce transitions out and new personnel come on board, should be a goal. The following traits and skills are considered essential for effectively carrying out the Reclamation mission in the years ahead:

- Integrated decision-making processes for assessment and management of risk and for the prioritization of projects.
- Integrated and expanded expertise for dealing with environmental, financial, social, legal, and resource conservation issues.
- Ability to work collaboratively with others, both within and outside the bureau.
- Clear, effective, and responsive communicators with sponsors, customers, contractors, Congress, state and local officials, tribal leadership, other governmental agencies, and the public.
- Technical, administrative, and management knowledge needed to define, assign, supervise, review, and evaluate outsourced work—people with such know-how are known as smart buyers.
- Technical and craft skills to accomplish inherently government functions that must be retained by Reclamation.
- Strong asset management skills for dealing with the operation, maintenance, rehabilitation, and replacement of aging infrastructure.
- Coordinated project management that incorporates continuous communication among all participants.
- Dedication to healthy research and development activities that focus on future needs and areas not duplicated by others.

Findings and Recommendations

Workforce and Human Resources

Finding 9a. Reclamation and other federal agencies recognize that successful outsourcing of technical services requires maintaining technical core competencies to develop contract scope, select contractors, and manage contracts, and that it is necessary for agency personnel to execute projects as well as to receive continuing training in order to maintain those competencies.

Finding 9b. Reclamation's current work is dominated by two categories of tasks: (1) the operation, maintenance, and rehabilitation of existing structures and systems and (2) the creation and brokering of agreements among a variety of groups and interests affected by the management of water resources. The need to include a broad spectrum of stakeholders, particularly groups that represent environmental issues and American Indian water rights, considerably alters both the tasks of the agency and the skills required to accomplish them.

Finding 9c. Reclamation employees appear on the whole to be more motivated by complex technical tasks than by tasks that are socially and politically complex. However, an increasing proportion of the work that employees at all levels engage in involves tasks that are socially and politically complex. Reclamation's current mission requires personnel to be equipped to address both technical uncertainties and the ambiguities of future social and environmental outcomes.

Recommendation 9a. Reclamation should do an analysis of the competencies required for its personnel to oversee and provide contract administration for outsourced activities. Training programs should ensure that those undertaking the functions of the contracting officer's technical representative are equipped to provide the appropriate oversight to ensure that Reclamation needs continue to be met as mission execution is transferred.

Recommendation 9b. In light of the large number of retirements projected over the next few years and the potential loss of institutional memory inherent in these retirements, a formal review should be conducted to determine what level of core capability should be maintained to ensure that Reclamation remains an effective and informed buyer of contracted services.

Recommendation 9c. Reclamation should recruit, train, and nurture personnel who have the skills needed to manage processes involving technical capabilities as well as communications and collaborative processes. Collaborative competencies should be systematically related to job categories and the processes of hiring, training, evaluating the performance of, and promoting employees.

Recommendation 9d. Reclamation should facilitate development of the skills needed for succeeding at socially and politically complex tasks by adapting and adopting a small-wins approach to organizing employee efforts and taking advantage of the opportunities to celebrate and build on successes.

ALTERNATIVE SCENARIOS FOR FUTURE INFRASTRUCTURE MANAGEMENT

The Nobel laureate physicist Nils Bohr once said that "prediction is very difficult, especially if it's about the future." However, the scenarios presented in this report are not predicting the future; they are only suggesting what is possible consistent with trends in workload and governmental mandates.

Finding and Recommendation

Finding 10. While the committee recognizes that the major changes suggested by the alternative scenarios are inappropriate for immediate implementation, the continuation and intensification of identified trends, as described in this report, could lead to a need for dramatic changes in Reclamation's operations and procedures in the years to come. The three future scenarios presented in this report—(1) a centrally located project management organization, (2) outsourced O&M, and (3) federal funding and local execution—provide a basis for anticipating future trends and preparing for future change.

Recommendation 10. Reclamation should consider the suggested future scenarios as a basis for analyzing longer-term trends and change.

Appendixes

Appendix A

Biographies of Committee Members

James Kenneth Mitchell (National Academy of Engineering and National Academy of Sciences), *Chair*, is University Distinguished Professor emeritus, Virginia Polytechnic Institute and State University, Blacksburg, Virginia, and a consulting geotechnical engineer. He was previously on the faculty of the University of California, Berkeley, from 1958 until his retirement as chair of the civil engineering department in 1993. His primary research activities focused on experimental and analytical studies of soil behavior related to geotechnical problems, including mitigation of ground failure risk during earthquakes. He has authored more than 350 publications, including guidance documents on soil stabilization, waste containment, ground improvement, and earth reinforcement, and a video, "Ground Improvement for Dam Safety," produced in 1998 by the Interagency Committee on Dam Safety. As a consultant, Dr. Mitchell has worked with numerous governmental and private organizations on geotechnical problems and earthwork projects of many types, especially soil stabilization, ground improvement for seismic risk mitigation, earthwork construction, and environmental geotechnology, both nationally and internationally. He is licensed as a civil engineer and as a geotechnical engineer in California and as a professional engineer in Virginia. He is a fellow and honorary member of the American Society of Civil Engineers. He served as secretary (1966-1969), vice chairman (1970), and chairman (1971) of the Geotechnical Engineering Division of ASCE and as chairman of the U.S. National Committee for the International Society for Soil Mechanics and Foundation Engineering. Dr. Mitchell was elected to membership in the National Academy of Engineering in 1976 and the National Academy of Sciences in 1998. He is the 2003-2005 chair of the Civil Engineering Section of the National Academy of Engineering. He has participated on 17 NRC boards and study committees and served as chair or vice chair of five. He has received numerous honors, including the Norman Medal in 1972 and 1995, the Thomas A. Middlebrooks Award (four times), the Walter L. Huber Research Prize, the Terzaghi Lecture Award, the Karl Terzaghi Award, and the H. Bolton Seed Medal (2004), all from the American Society of Civil Engineers, and the U.S. Army Corps of Engineers' Chief of Engineers Outstanding Service Award in 1999. Dr. Mitchell received a B.S. in civil engineering from Rensselaer Polytechnic Institute in 1951, an M.S. in civil engineering from the Massachusetts Institute of Technology (MIT) in 1953, and a Ph.D. in civil engineering, also from MIT, in 1956.

Patrick R. Atkins is director of environmental affairs at Alcoa, where he is responsible for environmental policy and global environmental programs. He serves on various lead teams, and he chairs global advisory committees that provide input to Alcoa's corporate environment, health, and safety programs. Dr. Atkins joined Alcoa in Pittsburgh in 1972, after having served as a professor in environmental health engineering at the University of Texas at Austin. He has published more than 50 technical articles and edited two books. Dr. Atkins is a member of the American Society of Civil Engineers, the National Society of Professional Engineers, and the Engineering Society of Western Pennsylvania. He represents Alcoa on the environmental committees of the International Primary Aluminum Institute, the Business Roundtable, the National Association of Manufacturers, and other national and international groups. In addition, he was a member of the National Academy of Sciences Commission on Geosciences, Environment, and Resources. Dr. Atkins is a registered professional engineer and an adjunct professor at the University of Pittsburgh Graduate School of Public Health, teaching industrial waste treatment technology. Dr. Atkins earned a B.S. in civil engineering from the University of Kentucky and an M.S. and Ph.D. in environmental engineering from Stanford University.

Allan V. Burman is president of Jefferson Solutions, a division of the Jefferson Consulting Group, a firm that provides change management services and acquisition reform training to many federal departments and agencies. Dr. Burman provides strategic consulting services to private sector firms doing business with the federal government as well as to federal agencies and other government entities. He also has advised firms, congressional committees, and federal and state agencies on a variety of management and acquisition reform matters. Prior to joining the Jefferson

Consulting Group, Dr. Burman had a long career in the federal government, including serving as administrator for federal procurement policy in the Office of Management and Budget (OMB), where he testified before Congress over 40 times on management, acquisition, and budget matters. Dr. Burman authored the 1991 policy letter that established performance-based contracting and greater reliance, where appropriate, on fixed-price contracting, as the favored approach for contract reform. As a member of the Senior Executive Service, Dr. Burman served as chief of the Air Force Branch in OMB's National Security Division and was the first OMB branch chief to receive a Presidential Rank Award. Dr. Burman is a fellow of the National Academy of Public Administration, a fellow and member of the board of advisors of the National Contract Management Association, a principal of the Council for Excellence in Government, a director of the Procurement Round Table, and an honorary member of the National Defense Industrial Association. From 1997 to 2003 he was a contributing editor and writer for *Government Executive* magazine. He has served as a member of the NRC Committee on Oversight and Assessment of Department of Energy Project Management since 2000. Dr. Burman obtained a B.A. from Wesleyan University, was a Fulbright scholar at the Institute of Political Studies, University of Bordeaux, France, and has a graduate degree from Harvard University and a Ph.D. from the George Washington University.

Timothy J. Connolly is senior vice president and a national director of quality at HDR Engineering, Inc. He is a professional structural engineer responsible for structuring project teams and restructuring poorly performing departments in HDR. He has led HDR's internal peer review program to review operational methods and procedures and develop an action plan to strengthen their effectiveness as well as identify the best practices that contribute to the success of the company. He is a member of the American Society of Civil Engineers, the American Railway Engineering and Maintenance Association, and the Society of American Military Engineers. He earned a B.S. and an M.S. in civil engineering from the University of Kansas.

Lloyd A. Duscha (National Academy of Engineering) retired from the U.S. Army Corps of Engineers in 1990 as the highest-ranking civilian after serving as deputy director, Engineering and Construction Directorate, at headquarters. He was principal investigator for the NRC report *Assessing the Need for Independent Project Reviews in the Department of Energy* and a member of the committee that produced the NRC report *Improving Project Management in the Department of Energy*. Mr. Duscha served in numerous progressive Army Corps of Engineers positions in various locations over four decades. Mr. Duscha is currently an engineering consultant to various national and foreign government agencies, the World Bank, and private sector clients. He served on the Committee on the Outsourcing of the Management of Planning, Design, and Construction Related Services as well as the Committee on Shore Installation Readiness and Management. He chaired the NRC Committee on Research Needs for Transuranic and Mixed Waste at Department of Energy Sites and serves on the Committee on Opportunities for Accelerating the Characterization and Treatment of Nuclear Waste. He has also served on the Board on Infrastructure and the Constructed Environment and was vice chairman of the U.S. National Committee on Tunneling Technology. Other positions held were president, U.S. Committee on Large Dams; chair, Committee on Dam Safety, International Commission on Large Dams; executive committee, Construction Industry Institute; and board of directors, Research and Management Foundation of the American Consulting Engineers Council. Mr. Duscha has numerous professional affiliations, including fellowships in the American Society of Civil Engineers and the Society of American Military Engineers. He holds a B.S. degree in civil engineering from the University of Minnesota, which awarded him the Board of Regents' Outstanding Achievement Award.

G. Brian Estes completed 30 years in the Navy Civil Engineering Corps, achieving the rank of rear admiral. Admiral Estes served as commander of the Pacific Division of the Naval Facilities Engineering Command and as commander of the Third Naval Construction Brigade at Pearl Harbor. He supervised over 700 engineers, 8,000 Seabees, and 4,000 other employees providing public works management, environmental support, family housing support, and facility planning, design, and construction services. As vice commander, Naval Facilities Engineering Command, Admiral Estes led the total quality management transformation at headquarters and two updates of the corporate strategic plan. He directed execution of the $2 billion military construction program and the $3 billion facilities management program while serving as deputy commander for facilities acquisition and deputy commander for public works, Naval Facilities Engineering Command. After retiring from the Navy he became the director of construction projects at Westinghouse Hanford Company, where he directed project management functions supporting operations and environmental cleanup of the Department of Energy's Hanford nuclear complex. He served on the committee that produced a series on progress in improving project management at the Department of Energy and has served on a number of other NRC committees. He holds a B.S. in civil engineering from the University of Maine, an M.S. in civil engineering from the University of Illinois, and is a registered professional engineer in Illinois and Virginia.

Martha S. Feldman is professor of planning, policy, and design management, political science, and sociology, and Roger W. and Janice M. Johnson Chair in Civic Governance and Public Management at the University of California, Irvine. She has a long-standing interest in how organizations influence people's ability to accomplish work. Her work in public management builds on this interest and focuses on the tools managers can use to create public organizations that are broadly inclusive of employees and the public. Prior to joining the University of California, she was professor of political science and public policy and associate dean of the Gerald R. Ford School of Public Policy at the University of Michigan. She has authored or coauthored four books, including *Strategies for Interpreting Qualitative Data* (1995), and scores of journal articles, book chapters, and reviews and commentaries. She has presented more than 40 papers, including one entitled "Organizational change process: Moving from plans to action" and "Organizational process and democratic capacity" at the Seventh National Public Management Research Conference (2003). Dr. Feldman is a member of the Academy of Management, the American Political Science Association, the American Society for Public Administration, and the Public Management Research Association. She holds a B.A. in political science from the University of Washington and M.A. and Ph.D. degrees from Stanford University.

Darrell G. Fontane is director of the International School for Water Resources and a professor in civil engineering at Colorado State University. His research interests include water resources decision support systems, computer-aided water management, and integrated water quantity and quality management. He is responsible for organizing international nondegree programs for engineers in various aspects of water resources engineering. Dr. Fontane served as a visiting associate professor at the Center for Water Resources and Quality Management, Korea, 1991, and as a visiting research engineer at the U.S. Army Corps of Engineers Waterways Experiment Station. He has served as principal or coprincipal investigator for research projects on topics such as methodologies to improve regional exchange of hydropower resources, stochastic analysis of project dependable capacity in hydropower systems, optimal design and operation of selective withdrawal structures, optimal selection of salinity control measures in the Colorado River Basin, developing alternative operation strategies for the Colorado River Basin, evaluation of the Lake Nasser optimization models, development of methods to assess alternative water-based recreational strategies, development of a general reservoir decision support system, and optimal operation of a system of lakes for quantity and quality. These projects have been funded by the World Bank, the U.S. Agency for International Development, the U.S. National Park Service, the U.S. Bureau of Reclamation, the U.S. Army Corps of Engineers, the U.S. Department of Energy–Western Area Power Administration, and the Korea Center for Water Resources and Quality Management. Dr. Fontane has served as a member of several NRC committees on issues related to water resources management, instream flows, and salmon survival in the Columbia River. He is a member of water resources professional societies such as the American Society of Civil Engineers (ASCE), the American Water Resources Association, and the International Water Resources Association. Dr. Fontane has over 95 publications, including several articles and presented papers on the analysis, planning, and management of water service systems for the ASCE. Dr. Fontane holds a B.S. in civil engineering from Louisiana State University, an M.S. from Georgia Institute of Technology, and a Ph.D. in civil engineering and water resources planning and management from Colorado State University.

Sammie D. Guy is a consulting engineer specializing in the prevention and resolution of disputes in the construction of water resource facilities. He retired from the Bureau of Reclamation after more than 30 years' service as an engineer and administrator. His positions included head of the construction contract branch, director for engineering research, and chief of international affairs, where he was responsible for providing technical assistance and training in water resources development and management to developing countries. He is a recipient of the Department of the Interior's Honor Award for Superior Service. Mr. Guy has also worked with the World Bank to provide technical assistance for construction management, quality assurance, and institutional organization in Indonesia and India. He is coauthor of a book on construction claims, now in its third edition, a member of the board of directors of the Dispute Resolution Board Foundation, the American Society of Civil Engineers (life member), the U.S. Committee on Irrigation and Drainage, the U.S. Committee on Large Dams, and the International Commission on Irrigation and Drainage. Mr. Guy holds B.S. and M.S. degrees in civil engineering from the University of Kentucky.

L. Michael Kaas recently retired as director of the Department of the Interior's Office of Managing Risk and Public Safety. In that position he was responsible for facilities management and health and safety. His 28-year career at the department also included positions at the U.S. Bureau of

Mines as associate director for information and analysis, chief of the Division of Resource Evaluation, chief of the Division of Environmental Technology Research, chief of the Office of Regulatory Projects Coordination, chief of the Division of Mineral Information Systems, deputy director of minerals information and analysis, and planning officer. He is a recipient of the Department of the Interior's Distinguished Service Award and its Meritorious Service Award. Mr. Kaas is a member and past director of the Society for Mining, Metallurgy, and Exploration (SME) of the American Institute of Mining, Metallurgical, and Petroleum Engineers (AIME) and a recipient of the Herbert Hoover Award. He has authored many technical papers. Mr. Kaas is a registered professional engineer in Minnesota and holds a B.S. in mining engineering from the Pennsylvania State University and an M.S. in mineral engineering from the University of Minnesota.

Charles I. McGinnis retired from the U.S. Army as a major general and was formerly the director of civil works for the U.S. Army Corps of Engineers. More recently he served in senior positions at the Construction Industry Institute in Austin, Texas. He has also served as a senior officer of Fru-Con Corporation and as the director of engineering and construction for the Panama Canal Company and later as vice president of the company and lieutenant governor of the Canal Zone. As director of civil works, he was responsible for a $3 billion per year planning, design, construction, operation, and maintenance program of water-resources-oriented public works on a nationwide basis. He is a fellow of the Society of American Military Engineers, a fellow and life member of the American Society of Civil Engineers, and a charter member of the National Academy of Construction. He is a recipient of the U.S. Army's Distinguished Service Medal. General McGinnis holds a master's degree in civil engineering from Texas A&M University.

Roger K. Patterson is a water resources consultant. He recently retired as the director of the Nebraska Department of Natural Resources. Prior to his appointment with the state of Nebraska, he spent 25 years with the Bureau of Reclamation working in several western states. He helped implement the Central Valley Project Improvement Act of 1992, landmark reform legislation involving more than 100 separate mandates that address project operations such as water conservation, contract renewals, and water transfers. A founding member of the Federal Ecosystem Directorate, Mr. Patterson was responsible for coordination among four federal agencies on issues related to protecting the San Francisco Bay and the Sacramento/San Joaquin River Delta. In 1995 he received the Presidential Rank Distinguished Executive Award for his leadership role in the development and supervision of water resources management programs in California and a Department of the Interior award for Distinguished Service. Mr. Patterson was chairman of the Nebraska Boundary Compact Commission and the state's representative to the Missouri River Basin Association, State Environmental Trust Board, Blue River Compact, Republican River Compact, and Upper Niobrara River Compact. He holds B.S. degrees in civil and environmental engineering from the University of Nebraska.

Appendix B

Briefings to the Committee and Discussions

OPEN COMMITTEE MEETINGS

February 28, 2005

Opening Comment
Tom Weimer, Department of the Interior, Water and Science, acting assistant secretary
John W. Keys III, Bureau of Reclamation, commissioner

History of Reclamation
Brit Storey, Bureau of Reclamation, Office of Program & Policy Services, senior historian

Reclamation Today—John Keys and selected staff
Mark Limbaugh, Bureau of Reclamation, director, external and intergovernmental affairs and deputy commissioner
Bill Rinne, Bureau of Reclamation, director of operations, and deputy commissioner
Bob Wolf, Bureau of Reclamation, director of program and budget

- Organization
- Reclamation role, core mission, and self image
- Reclamation budget and factors that determine the budget
- Overview of Reclamation facilities and infrastructure
- Major construction

- Relationships with stakeholders for water and power (other federal agencies, Congress, environmental groups, public interest groups, and states)
- High-profile issues

April 6-7, 2005

Welcome and Introductions
Fred Ore, Operations, deputy director

Delivering Water and Generating Power
Robert Johnson, Lower Colorado Region, regional director
Brian Person, Eastern Colorado Area Office, area manager

Security Safety and Law Enforcement
Larry Todd, Security Safety and Law Enforcement, director
Bruce Muller, Dam Safety Office, chief

Policy Management and Technical Services
Michael Gabaldon, Policy, Management, and Technical Services (PMTS), director

Technical Service Center
Michael Roluti, Technical Service Center (TSC), director

Bureau of Reclamation Laboratory tour and discussion
Michael Roluti, TSC, director
Cliff Pugh, Water Resources Research Laboratory Group, manager

Project Cost Overview (flow of money)
Ephraim Escalante, Finance and Accounting System, manager

Administrative Requirements (Centralized Management, A-76)
Elizabeth Harrison, Management Services Office, director

Acquisition and Contracting
Karla Smiley, Acquisition and Assistance, manager

Roundtable Discussions on Case Studies and Site Visits of the Colorado–Big Thompson Project
Brian Person, Eastern Colorado Area Office, manager
Mike Applegate, Northern Colorado Water Conservancy District, president

Eric Wilkinson, Northern Colorado Water Conservancy District, general manager

June 22-24, 2005

Roundtable Discussion to Determine Organizational and Operating Models and to Identify Good Practice Tools and Techniques for Infrastructure Management
Donald Basham, U.S. Army Corps of Engineers, Engineering and Construction, chief
Janet C. Herrin, Tennessee Valley Authority, River Operations, senior vice president
Leslie F. Harder, California Department of Water Resources, Division of Flood Management, director

Reclamation Customer Roundtable Discussion on Reclamation Strengths, Weaknesses, Opportunities, and Threats
Dan Keppen, Family Farm Alliance
Tom Donnelly, National Water Resources Association

Discussions with Senate Energy and Natural Resources Committee Staff
Kellie A. Donnelly
Michael L. Connor
Nathan Gentry

August 16, 2005

Roundtable Discussion of Environmental Issues that Affect the Design, Construction, Operation, and Maintenance of Reclamation Facilities and Infrastructure and the Bureau's Organization
Thomas J. Graff, Environmental Defense, regional director

COMMITTEE DISCUSSIONS AND SITES VISITED AT RECLAMATION REGIONS

Two- to three-member delegations from the committee visited Reclamation regions between April 8, 2005, and June 10, 2005. The visits involved meetings with regional office managers; regional division managers for the environment, operations and maintenance, construction, engineering design, planning, contracting and finance, and human resources; area office and project managers; and representatives of Reclamation power and water customers and contractors. The meetings addressed discussion questions (listed below) developed by the committee, but they were loosely structured to encourage a free exchange of ideas and opinions. The meetings also provided the committee an opportunity to ask follow-up questions regarding Reclamation's written response to the committee's request for background data (listed below).

Meetings were conducted with the following Reclamation offices and customer organizations:

Animas–La Plata Project Office
Boise Board of Control
Lower Colorado Dams Office
Central Utah Project
Central Valley Project Water Association
Colorado River Commission-Nevada
Eastern Colorado Area Office
Lower Colorado Regional Office
Mid-Pacific Regional Office
Northern California Power Agency
Pacific Northwest Regional Office
Provo Area Office
Provo Water District
San Juan Water Commission
San Luis and Delta Mendota Water Authority
Snake River Area Office
Upper Colorado Regional Office
Upper Colorado area offices
Utah Reclamation Mitigation and Conservation Commission

The sites visited included the following facilities:

Arrowrock Dam
Boise Diversion
Davis Dam
Deer Creek Dam
Hoover Dam
Jordanelle Dam environmental restoration
Parker Dam
Tracey Fish Collection Facility
Tracey Pump Facility

Meetings were conducted with the following organizations via conference calls:

Bonneville Power Authority
Colorado River Energy Distributors Association
Great Plains Regional Office
Navajo Nation
Southern Ute Department of Natural Resources

DISCUSSION QUESTIONS

The following questions were used to guide informal discussions between Reclamation personnel and committee site visit groups and between Reclamation customers and contractors and the committee site visit groups.

Overarching question
What do you see changing over the next 5, 10, 25 years, and what will you need to do to address these issues?

1. How do you rate the performance of the Reclamation Technical Service Center on a scale of 1 to 10, with 10 being excellent and 1 being totally unacceptable?
   a. responsiveness,
   b. quality of service, and
   c. cost

2. What is your interpretation of Reclamation's mission, and how does it apply to the work in your region/area?

3. What do you see as the greatest obstacle to achieving your mission now and in the future?

4. If you could change one Reclamation policy or requirement, what would it be and how would you change it?

5. What additional engineering and construction activity do you think your office could absorb effectively and easily?

6. What do you see other organizations (public and private) doing that if adopted by Reclamation would make your job easier?

BACKGROUND DATA QUESTIONS

The regional offices were requested to provide written responses to the following questions:

Human Resources

1. (a) How many employees are there in your regional and area offices? (b) What are Reclamation's personnel resources and how are they distributed in the regional and area offices (percentages of staff are more important than actual numbers)?

Location	Design	Other Technical	Construction	O&M	Management	Support	Legal
Regional offices							
Area offices 1							
Area offices 2							
Area offices etc.							

2. What disciplines and specialties are included under "Other Technical" personnel (e.g., biologists)? Where are these disciplines located?

3. Are personnel allocated according to mission elements (power, water, other operations) or are the same technical experts available for all the mission elements?

4. What are the major differences in required skills and technologies for building dams versus rehabilitating or rebuilding them?

5. Do regional and area offices have the personnel resources (numbers and skills) needed to undertake the mission now? In the future? If shortages exist, what skills and in what specific areas?

6. What difficulties, if any, have the regional and area offices faced in recruiting personnel with the required engineering or other technical expertise?

7. What percentage of staff is projected to retire in the next 5 years? What skills will they represent? How might this affect the future composition of the workforce? What strategies are in place to retain staff? to recruit new talent?

APPENDIX B

8. What personnel career development and training programs do the regional and or area offices have in place? How are they funded and at what level? How are these programs implemented? How are the staff who participate in these programs recognized and rewarded?

9. What processes or systems are in place to capture the regional and area offices' institutional memory?

Workload

Location	Number of Projects	Number of Irrigation Facilities	Number of Power Facilities	O&M Budget	O&M Backlog	Number of Construction Projects	Value of Constr. Projects
Area offices 1							
Area offices 2							
Area offices etc.							
Total for region							

10. What are the critical issues regarding execution of the workload? What is the projected workload for the next 5-10 years?

11. What are the critical issues regarding compliance with regulatory responsibilities (e.g., the 1982 Reclamation Act, the Endangered Species Act, Native American water rights)?

12. What impacts have requirements for increased security had on the workload, budget, personnel allocation, and methods of operation?

13. Are there any elements of the current workload that are decreasing and could go away in the future? Are there any anticipated new elements?

14. How are operations and maintenance activities and costs changing as the infrastructure ages?

Contracting Environment

15. What services/functions are currently being outsourced? How are these services/functions distributed—that is, is there greater use of outsourcing in some areas than others? If so, what might be driving these differences?

16. What core competencies (knowledge, skills, and abilities) are required in-house for Reclamation to effectively manage outsourced activities? Are these skills available now?

17. Does Reclamation measure the results/performance of its outsourced activities? If so, how?

18. Given that regional and area offices have the option of using Reclamation's Technical Service Center or outsourcing, what are the historical trends? What reasons have been given for selecting one option or the other?

19. What projects and activities include customer pay-for-service and co-pay of expenses? How are they included in the budget? What are the mechanisms for repayment?

Asset Management

20. How are projects currently managed (as a portfolio, regionally, for river basins, or as individual entities)? Are there any plans to change current management practices? If so, what are they? What is driving the changes? What outcomes are expected?

21. What decision-making processes and procedures are used to prioritize construction projects? O&M activities? Is there documentation for these processes/procedures?

22. Does Reclamation apply adaptive management techniques? What has been the experience?

23. What types of internal and external reviews (management and technical) are routinely conducted and how are the results used?

24. What performance measures are used for asset management?

25. What internal or external benchmarking activities are undertaken?

Operations

26. What are the regional and area office relationships with other organizations, including the Army Corps of Engineers, Bonneville Power, Western Area Power Administration, Nature Conservancy, Natural Resources Defense Council, Western Governors' Association, Council of State Governments West? Others of import?

27. How smooth are the working relationships between TSC, the regions, and the area offices? What works well? What doesn't? What are your suggestions for improvement?

Construction

28. Are construction project management policies and procedures from inception through preproject planning, design, construction, and commissioning determined by Denver or by the regional or area offices?

29. How are construction project teams structured (types of expertise; in-house staff or contractors)?

30. How are accountability and responsibility assigned? Who signs off on a project? Who is responsible for any failures? Who has administrative and technical responsibility? How is performance assessed?

31. What contracting and delivery methods are used for construction projects? Are any new methods being considered for future use? If so, what training might be required?

Research and International Activities

32. What research activities are undertaken at the regional or area office to exchange/gather information on issues of science and technology?

33. What other issues, challenges, operating procedures should the committee be aware of in conducting this study?

Appendix C

Good Practice Tools and Techniques Roundtable

On June 22, 2005, the committee convened a meeting to discuss organizational and operating models used by other federal agencies and other governmental organizations with mission responsibilities similar to those of Reclamation to identify good practice tools and techniques. Representatives of the U.S. Army Corps of Engineers (USACE), the Tennessee Valley Authority (TVA), and the California Department of Water Resources (DWR) participated in the discussion.[1] The focus of the discussion was the facility development and management practices used by these organizations. More specifically,

- What expertise is needed to develop and manage facilities and infrastructure?
- Are human resources functions centralized or decentralized?
- How autonomous are regional and subregional offices in setting policies and procedures and making facility and infrastructure decisions?
- How are policy and procedures developed and documented?
- How are engineering services organized and provided?
- What is the impact of environmental requirements and how are they addressed?

---

[1] Guests included Donald Basham, chief, Engineering and Construction, U.S. Army Corps of Engineers; Janet Herrin, senior vice president, River Operations, Tennessee Valley Authority; and Leslie Harder, director, Division of Flood Management, California Department of Water Resources.

- How and when are engineering services outsourced?
- How are budgets for facilities and infrastructure developed and what are the sources of funding?
- How are customers involved in the budget planning process?

U.S. ARMY CORPS OF ENGINEERS

USACE's civil works mission is very similar to Reclamation's. The main difference is that Reclamation's operations are focused in the western states and USACE operates throughout the country. Reclamation has more of a focus on providing water for irrigation and USACE has a greater emphasis on flood control and navigation. Both organizations have had major construction programs to develop dams and waterways and are now responsible for the operation, maintenance, and recapitalization of these facilities.

USACE is composed of 41 districts each having a fairly high degree of autonomy. The districts are organized into 8 regions. Current mission requirements are driving USACE toward more uniform policies, procedures, and service to customers by reducing autonomy. For example, in order to move drawings and plans among regions, the format and nomenclature of computer-assisted design and drafting applications need to be 100 percent consistent. Pressure to downsize the organization means that USACE may not be able to have all disciplines and expertise needed in every district, which will result in shifting of work and personnel. The movement of work and personnel within the organization will require consistent policies and procedures to work effectively and to avoid instituting reductions in force in one area while simultaneously increasing staff in another. The prevalence of family structures with two wage earners makes it more difficult to geographically move people to implement reorganization. Standardized procedures facilitate the organization's capability to work together from dispersed locations.

Headquarters staff has shrunk from about 1,500 to about 750 today. This reduction has been accomplished by shifting responsibilities to the field. For example, policy is now developed by communities of practice in the field rather than by permanent headquarters staff. This has the advantage of having policy developed by the people who will have to implement it.

Personnel recruiting is generally decentralized, although many regional office are assuming a greater role in order to balance staffing and workload across the region. The recruitment and selection of regional division chiefs in each district (e.g., chiefs for real estate, planning, engineering, and construction divisions) is undertaken with the personal involvement of the respective headquarters discipline chiefs. This is done to ensure consistency throughout the corps. The whole point of USACE is to have technical competency, but at a certain level of the organization, technically competent staff is not sufficient. The corps also needs people with strong leadership capabilities.

USACE uses centralized guidance with local implementation. Projects are developed locally by district offices that interact with the sponsors and other local stakeholders. Projects are developed by teams with the necessary technical expertise, which may include the construction trades, engineering, botany, biology, social sciences, economics, resource management, project management, and other kinds of expertise needed to undertake the complex and varied projects assigned to the corps. This works better than projects undertaken by discipline stovepipes (e.g., planning, engineering, and construction) that do their work then pass the project on to another discipline. There is an increasing emphasis on ensuring that the people with technical expertise also have leadership skills. This is accomplished through career development and training programs for technical personnel.

Project management plans are developed at the beginning of projects. The sponsors play a significant role in developing the project scope and execution plans. Sponsors also participate in contractor selection panels. Some more sophisticated sponsors participate in the design process.

USACE also relies on contractors to achieve its mission. All construction work is contracted. Seventy-five percent of the engineering and architecture for military construction is undertaken by contractors. USACE believes that it needs to undertake 25 percent of the work in-house in order to maintain the expertise necessary to effectively select and oversee contractors. In the last 10 years the percentage of in-house engineering for civil work dropped from 95 percent to about 54 percent. This drop is due in part to fluctuations in workload as well as to a reduction in the number of federal employees. USACE is in the process of undertaking an A-76 review and competitions for information technology (IT) and civilian works operations and maintenance (O&M). The IT initiative is being undertaken as a single contract for the entire corps so that regardless of the outcome, IT services will be more uniform across the agency. O&M contracts will be site-specific.

USACE does not have a central organization for technical expertise such as Reclamation's TSC. Most of USACE's design work is done in the district offices. Many senior engineers are located in headquarters, but the corps relies on a matrix of centers of expertise at the regional offices that provide services for all districts USACE-wide. For example, there is a hydropower design center in Portland that does all such work or reviews hydropower work undertaken by the districts. Current pressures in military construction are to strengthen the regional offices, and this is likely to follow for civil works.

The corps's five laboratories are now operated as a single lab system with headquarters at Vicksburg, Mississippi. Research is funded through military and civil works projects. There is also some direct funding for more basic research. Work is also undertaken in cooperation with universities. To some extent, the labs are competitive with those in the private sector, but for the most part they have unique, world-class capabilities. The corps partners with Reclamation for some research, although the level of this cooperation has diminished in recent years.

USACE has developed environmental operating principles. The current approach includes environmental consideration from the beginning of design. This approach may add to the first cost of the project but saves money in the long run. Project sponsors who pay a portion of the total costs sometimes resist including environmental mitigation features that increase the costs, but the corps helps them understand that this is part of the current method of executing projects.

USACE tries to use innovative contracting approaches within the bounds of federal regulations. Overall, about 40 percent of USACE construction is now design-build—more so for military construction than for civil works projects. An advantage is achieved in being able to overlap design and construction schedules. There is still some question about the extent to which this can be done on dams and related facilities. The limited number of contractors in this arena is also a factor. The corps also uses a lot of indefinite delivery/indefinite quantity contracts that help build long-term partnering relationships with contractors.

USACE develops its annual budget much like any other federal agency. In the end, the appropriated budget is about 80 percent proposed by the administration and 20 percent is added on by Congress. The current civil works budget is about $5 billion. All projects are undertaken with appropriated funds, but they are not implemented until sponsors secure their matching contributions where necessary.

TENNESSEE VALLEY AUTHORITY

TVA's mission is to generate prosperity for the Tennessee Valley. There are three goals: provide low cost, reliable power, support a thriving river system and environment, and support economic development. These goals encompass requirements for maintaining navigation and flood control, established in the initial TVA legislation. TVA is both a power producer and a power marketer and operates as a federal corporation.

TVA has an annual budget of over $7 billion and is the nation's largest public power provider, serving 8.5 million residents and 650,000 businesses and industries. In addition to its ratepayers, the TVA has many public and private stakeholders that are affected by how TVA manages the Tennessee River and TVA facilities and infrastructure. TVA is fully funded by its ratepayers; it has not received any federal appropriations since 1999. As a regional natural resource manager, TVA sells power to 158 local distributors and serves 62 industrial and federal customers directly. TVA has about 33,000 megawatts of capacity with a mix of hydro, coal, nuclear, wind, solar, and methane power generation.

At the direction of Congress, TVA is currently transitioning its organization from management by a three-member, full-time board of directors and a chief operating officer to a nine-member, part-time board with a chief executive officer appointed by the board. It is too soon to determine how this change will affect operations.

TVA operates in a 40,000-square-mile watershed and provides electricity to an 80,000-square-mile service area. The incongruity of the environmental impact area and the service area means that there are some ratepayers who are not stakeholders. TVA is the watershed manager and has congressionally mandated environmental stewardship functions whose costs are part of the operating expenses and are included in the rates charged for power. The 12 watersheds that feed the TVA dam and reservoir system are managed by teams that work with local stakeholders to control erosion and maintain water quality.

TVA works closely with USACE headquarters and its Cincinnati-based regional office. Legislation that established TVA makes it responsible first for flood control and navigation, which must be met before generation of electricity. TVA owns locks that are operated by USACE, requiring a close relationship to plan, construct, maintain, and rehabilitate the lock system. TVA also provides assistance to USACE for modernization of USACE generating facilities.

TVA's River System Operations and Environment group is organized into five functional units, including resource stewardship; environmental policy and planning; research and technology applications; river operations; and business services. Administrative functions, such as human resources, are centralized.

TVA has just over 12,000 employees, down from about 50,000 in the 1980s. It moved out of the construction business in 1988, resulting in a massive reduction in force. TVA staff are federal employees but not part of the civil service. All design and construction is now undertaken by contractors that have long-term partnerships with TVA.

River Operations has an annual budget of about $170 million. About half of the budget is for O&M and about half is recapitalization of the aging infrastructure. The average age of TVA facilities is 65 to 70 years.

Many of the units have not been modernized since they were first constructed. A power train modernization program was started in 1991. Fifty-one units have been updated; 41 units are still to be modernized, with completion scheduled for 2015. This is the only major capital program within River Operations. The program has been and can still be the target of budget reductions that extend the schedule. The extended recapitalization schedule results in additional O&M requirements to keep the units operating. Staff recommend project priorities based on broad budgetary constraints set by the TVA board.

About half of the employees in River Operations are skilled craftspeople, including electricians, machinists, and operators. In the last 3 or 4 years there has been a transition to multiskilled craftpeople to increase staff efficiency in a more automated environment. The transition involved a reduction in force through attrition, retraining of the existing workforce, and hiring of new multiskilled employees. New multiskilled hydro technicians receive about 30 months of training. The transition has not been without problems, but it has been aided by having detailed procedures in place. The workforce can be shifted across TVA to accommodate fluctuations in requirements. The modernization program is being undertaken by contractors under blanket agreements who bid on individual tasks.

As noted above, River Operations employs about 300 craftspeople. There are about 200 engineers, and the others are administrative. Civil, electrical, and mechanical engineers are in a central design group. Engineering support can also be obtained from other divisions—for example, cable engineering from the nuclear power division—or outsourced. There are also some contract employees who operate under TVA supervision. All TVA design engineering is based in Chattanooga. The water resources engineers, e.g., hydrologists, who operate the river system are located at the forecast center, which operates 24/7. Automation allows TVA hydro plants to be staffed 8/5.

Outsourcing decisions are based on availability and economic factors. TVA has developed outsourcing procedures consistent with union agreements for craftspeople and engineers. TVA is also shifting inventory requirements to suppliers and contractors.

TVA undertakes cooperative research programs with the Department of Energy's Oak Ridge National Laboratory (ORNL) in energy and water resource development. Some of the research is funded by TVA and some through grants from other sources. ORNL often uses TVA facilities for demonstration projects, which provide benefits at no cost to TVA.

At one time, TVA undertook international marketing of its expertise. The current policy is to respond to international requests for assistance when funding is provided.

TVA has the problem of an aging workforce, with missing generations in the middle. This makes succession planning and the maintenance of corporate knowledge difficult. TVA has initiated an engineering and scientific graduate progression program that outlines a developmental progression with on-the-job training and course requirements. The training imparts a combination of general and discipline-specific information, as well as TVA-specific procedures. An internal board determines when personnel are ready to progress to a higher level. The procedure was developed in conjunction with the engineer's union.

TVA recently undertook a comprehensive study with its stakeholders (federal, state, business, recreation, environmental, and natural resource organizations) to set priorities and revise reservoir operating plans. All aspects of reservoir operations were put on the table for the stakeholders, many having divergent priorities, to assess and make recommendations. Beginning with TVA's mandate for navigation and flood control, the stakeholders addressed the various recreational, environmental, and economic interests to develop operational priorities. After 2 years, a comprehensive operation plan was developed with the support of TVA and all its stakeholders. This plan redistributes both the risks and benefits of river system operations. TVA is conducting extensive monitoring to determine the effects of the new policy and will make adjustments if unexpected, unacceptable impacts are identified.

TVA undertakes some innovative contracting, such as performance-based contracts that link fees to schedule, safety, environmental, and other specific outcomes. The application is generally for large contracts that are used across TVA. Because the river operations aspect of these contracts is relatively small, the achievement of River Operations performance measures typically has minimal effect on overall contract performance.

CALIFORNIA DEPARTMENT OF WATER RESOURCES

DWR has about 2,500 employees, which is smaller than USACE and TVA. DWR has different constraints but faces many of the same issues. DWR's mission is to manage the water resources of California in collaboration with others to benefit the state's people and to protect, restore, and enhance the natural and human environment. Over 50 percent of DWR's personnel are assigned to the State Water Project (SWP), which covers much of the same geographic area as Reclamation's Central Valley Project (CVP) but is smaller and serves more urban customers. SWP includes 17 pumping plants, 8 hydroelectric plants, 30 storage facilities, and 693 miles of canals and pipelines. Energy requirements to pump water in the project make it the state's largest energy consumer. It is also the fourth largest power producer in the state.

APPENDIX C   135

Through its safety of dams program, DWR is also a public safety and regulatory agency responsible for 1,250 dams in the state. The safety of dams program also performs an oversight role in new construction. The Division of Flood Management is a public safety agency focused in the Central Valley as the nonfederal sponsor for federal flood control projects. DWR also provides water resource planning assistance to local governments and administers statewide electricity contracts.

DWR staff has a wide variety of expertise, giving it broad capabilities to address water management issues. Like other organizations, maintaining this expertise is a growing problem. Part of the reason is that the large construction projects that supported development of the expertise are no longer being undertaken. The last large project, the coastal aqueduct, was in the mid-1990s. Another is attrition of older, experienced personnel. The average experience of a typical DWR journeyman engineer has gone from about 20 years to a little over 2 years. Experienced personnel are brought in on a contract basis to work with DWR staff. This is effective in helping to mentor and train younger engineers. Consulting engineers are also employed for specific expertise and for design review boards.

It has been difficult to adapt personnel classifications and staffing levels as the organization transits from design and construction to O&M. There are very few nonengineers at the management level because engineers can migrate from technical areas to nontechnical areas to obtain promotions but nontechnical personnel cannot migrate to technical areas.

DWR now does between $30 and $100 million in construction work a year. All construction work is undertaken by private contractors. The contracts are administered and the work inspected by DWR staff. DWR has a small soils and concrete laboratory for construction support, because work undertaken by private laboratories has turned out to be of poor quality.

DWR is part of the state civil service system, which has a centralized personnel office. The rigidity of civil service regulations makes maintaining the necessary core competencies more difficult. DWR will need to find new ways to work with the system and have more flexible approaches, such as matrix management, to address these issues.

There are five field divisions with approximately 100 employees each to operate and maintain the system. There are also four districts for water resource planning and local government assistance. There is a centralized control center that can operate the whole project, but operations are still conducted at the division offices. The control center is located in the same building as Reclamation's Central Valley Project control center. The trend is toward centralization.

Each field division has a group of about 10 engineers who troubleshoot problems and direct O&M efforts. The headquarters engineering group works on review of proposed encroachments, corrosion analysis, and safety of dams issues. There is a centralized design and construction group with about 200 engineers, architects, cost estimators, and specification writers. Almost all of the design work for DWR's $30 million annual construction budget is done in-house. The organization has limited contracting ability, and state regulations preclude the application of design-build projects. Personnel in the field act as the owners of the facilities and as customers for project services, suggesting alternative solutions to the engineering problems but relying on engineers from the central design centers for project designs. This assures that the end product works for the facility operators.

The era of new large projects in California is over. New projects will be in the form of system modifications and improvement. These projects may require even greater planning, engineering, and construction skills than building new dams, and they will require a significant capital investment. The intertie between Reclamation's Delta-Mendota canal and DWR's Central Valley canal is an example of this type of project and of the increasing need for Reclamation and DWR to coordinate efforts to manage water in California. This project builds on established relationships at the interconnected and joint-use facilities in the system.

DWR funds are from water contractors who repay bonds and O&M and engineering expenses and who play a role in overseeing DWR's O&M activities. Convincing customers that they are being charged a fair share of the costs can be difficult at times. A 5-year strategic plan is the vehicle that expresses the need and timing for recapitalization projects.

DWR is starting to do some benchmarking of operations as well as of design and construction costs. Finding a benchmarking partner with a similar structure and processes and identifying appropriate metrics is difficult. Power and pumping plant operations seem to be much more amenable to benchmarking than do irrigation operations.

DWR addresses many of the same environmental issues addressed by Reclamation, such as counts of endangered fish species in the Sacramento Delta. Sustainability is becoming more important, and environmental issues are being incorporated in all design decisions. Environmental design expertise is generally provided by consultants. DWR is part of the CALFED process,[2] and many of the issues apply to DWR operations.

---

[2]The California Bay–Delta Authority oversees the implementation of the CALFED Bay–Delta Program for the 25 state and federal agencies working cooperatively to improve the quality and reliability of California's water supplies while restoring the San Francisco Bay–Sacramento/San Joaquin Delta ecosystem. The California Bay–Delta Act of 2003 established the authority as the new governance structure and charged it with providing accountability, ensuring balanced implementation, tracking and assessing program progress, using sound science, assuring public involvement and outreach, and coordinating and integrating related government programs.

Another environmental concern is the elderberry beetle, which lives in elderberry bushes. Conservation measures include a no-cut exclusion zone 100 feet around each elderberry bush. There are lots of elderberry bushes along waterways, making maintenance very difficult if not impossible. To address these issues, DWR is convening a series of interagency workshops with USACE, the U.S. Fish and Wildlife Service, the Water Quality Control Board, and others to develop short- and long-term solutions.

DWR, as the sponsor for the federal flood control project, is responsible for 1,600 miles of levees. It inspects the levees, and except for about 200 miles it maintains itself, relies on local Reclamation districts to undertake the needed maintenance. It is also responsible for the bypass channels and weirs that operate the channel system. It is faced with an infrastructure more than 50 years old and growing maintenance costs. In many areas, maintenance costs exceed initial construction costs. Maintenance costs and potential liability have significant impacts on the state's general fund. California is considering creation of a flood control authority that can assess beneficiaries of the system fees to cover the cost of maintenance and the cost of the potential liability.

Board on Infrastructure and the Constructed Environment

The Board on Infrastructure and the Constructed Environment (BICE) was established by the National Research Council (NRC) in 1946 as the Building Research Advisory Board. BICE and its predecessor organizations have been the principal units of the NRC concerned with the relationship between the constructed and natural environments and their interaction with human activities. Principal areas of focus include these:

- Human factors and the built environment,
- Project management methods,
- Construction methods and materials,
- Security of facilities and critical infrastructure,
- Multi-hazard mitigation methods,
- Construction and utilization of underground space, and
- Infrastructure and community building.

BICE brings together experts from a wide range of scientific, engineering, and social science disciplines to discuss potential studies of interest; develop and frame study tasks; ensure proper project planning; suggest possible reviewers for reports produced by fully independent ad hoc study committees; and convene meetings to examine strategic issues. The board members listed in the front of this document were not asked to endorse the committee's conclusions or recommendations, nor did they see the final draft of this report before its release.

Additional information about BICE can be obtained online at http://www.nationalacademies.org/bice.